Beginning Google SketchUp for 3D Printing

Sandeep Singh

Beginning Google SketchUp for 3D Printing

ISBN-13 (pbk): 978-1-4302-3361-9

ISBN-13 (electronic): 978-1-4302-3362-6

Printed and bound in the United States of America (POD)

Publisher and President: Paul Manning
Acquisitions Editor: Frank Pohlmann
Lead Editor: James Markham
Technical Reviewer: Michael Fredrickson
Editorial Board: Steve Anglin, Mark Beckner, Ewan Buckingham, Gary Cornell, Jonathan Gennick, Jonathan Hassell, Michelle Lowman, Matthew Moodie, Duncan Parkes, Jeffrey Pepper, Frank Pohlmann, Douglas Pundick, Ben Renow-Clarke, Dominic Shakeshaft, Matt Wade, Tom Welsh
Coordinating Editor: Candace English
Copy Editor: Kim Wimpsett
Compositor: MacPS, LLC
Indexer: Toma Mulligan
Cover Designer: Anna Ishchenko

Distributed to the book trade worldwide by Springer Science+Business Media, LLC., 233 Spring Street, 6th Floor, New York, NY 10013. Phone 1-800-SPRINGER, fax (201) 348-4505, e-mail orders-ny@springer-sbm.com, or visit www.springeronline.com.

For information on translations, please e-mail rights@apress.com, or visit www.apress.com.

Apress and friends of ED books may be purchased in bulk for academic, corporate, or promotional use. eBook versions and licenses are also available for most titles. For more information, reference our Special Bulk Sales–eBook Licensing web page at www.apress.com/info/bulksales.

To my parents, Baldev and Harjit; my wife, Khuspreet; and my sister, Balveen, for their love and support.

Contents at a Glance

Contents

About the Author

 **Sandeep Singh** is 27 years old and lives in Sacramento, California, with his wife. He currently works as a project engineer providing technical support for Energy Management Software (EMS). He has a bachelor's degree in electrical and electronics engineering from California State University, Sacramento and a master's degree in biomedical engineering from Cal Poly, San Luis Obispo. Sandeep started using SketchUp in 2007 after taking a class in product design and development where he learned how CAD software can be used to prototype models for biomedical prostheses and artificial organs. He found SketchUp to be the simplest modeling software to use and yet advanced enough to design complicated models for his own projects.

When not working on his projects, he enjoys reading how-to books, building things, running, hiking, watching movies, and spending time with his wife.

About the Technical Reviewer

Steve Nieman is a systems administrator/teacher's assistant for a private school in Munising, Michigan. He has earned degrees in accounting and information systems and has completed a course in PC repair. Prior to his current job, Steve enjoyed the privilege of being a stay-at-home dad and raising his three kids.

Acknowledgments

This book would not have been complete without the hard work of a lot of people. I would first like to thank the fine people at Apress for their support and guidance, especially Candace English, Frank Pohlmann, James Markham, and Jonathan Gennick, who answered many of my e-mails when writing this book. They are great people to work with. Our copy editor, Kim Wimpsett, was so very helpful in making sure the quality of the book was up to standard. A great thanks to Matthew Moodie and Douglas Pundick who jumped in toward the end to keep the review process following smoothly.

I would like to also thank the tech reviewer, Steve Nieman, who spent many hours going through each chapter and providing comments for improvement. And a big thank-you to Karen Embry, who prepared the proposal letter, reviewed the very first draft of the book, and suggested I submit the book for publication.

Lastly, I would like to thank my parents, Baldev Singh and Harjit K. Singh; my wife, Khuspreet Singh; and sister, Balveen K. Singh, for their love and support while I was writing this book.

Introduction

Welcome to *Beginning Google SketchUp for 3D Printing*. I'm pleased that you chose this book as a companion to guide you in your 3D printing and SketchUp modeling adventures. Whether you're an experienced pro or a seasoned 3D designer, hobbyist, craftsperson, or artist who is new to Google SketchUp and Shapeways, you will find valuable information and step-by-step instructions to help you develop your SketchUp expertise for 3D printing.

The book is divided into three sections. If you are a Google SketchUp and Shapeways beginner, I recommend you start with a close reading of Part I, because it lays out some of the basics of modeling in SketchUp and 3D printing in Shapeways. Those of you who are intermediate to advanced users can skim through the chapters to find what most appeals to you. Part II of the book focuses on developing models in Google SketchUp for 3D printing in Shapeways. Part III is all about presenting and sharing your models. The book concludes with future developments in 3D printing. To set the stage for you to start thinking about 3D printing, I'll introduce the different printing methods for some historical perspective.

Background

Advanced printing technology has been around for many decades now and has become highly advanced since the first printing presses were invented in the 15th century. As the years have progressed, we have seen changes in printing technology, and it's now available in various forms. Table 1 lists the five printing methods that are being used today.

Table 1. Five Common Printing Methods

Methods	Descriptions
Screen printing	This form of printing is commonly used to create designs on fabric. Screen printing involves a screen or wire mesh that paint seeps through to create a design. Most designs on T-shirts are screen printed.
Dot-matrix printing	This form of printing was commonly used before the invention of laser or inkjet printers. This printer is slower and nosier than an inkjet or laser printer. The printouts are created from dots. The print head strikes an ink ribbon that transfers a dot of ink onto paper. The alignment of these dots creates the characters on paper.
Inkjet printing	This is the type of printing most commonly used with at-home desktop printers. Desktop printers are relatively inexpensive and can be bought at your local electronics store. Expect to spend anywhere between $30 and $100 for an inkjet printer.

Methods	Descriptions
Laser printing	Laser printers are commonly found in a business setting. The benefits of laser printers are less noise and quick printing, but they are more expensive than the inkjet printer. Expect to spend between $150 and $400 for a desktop laser printer.
3D printing	One of the most advanced printing technologies involves 3D printing. There are several types of 3D printers, including nano printers for printing on the nano scale. There are 3D printers that print various types of materials. You will find 3D printers that use plastic, powder, and paper—and even chocolate and sugar. The possibilities are endless when you think about the types of materials you can use for 3D printing. 3D printers are available in all shapes and sizes. Some take up an entire room, and some can be easily placed on your desktop.

I hope that after reading this book you are encouraged to dive into your own fabricating adventures and try the many new ways of using SketchUp and Shapeways in your daily life.

Who This Book Is For

This book is for anyone interested in learning about 3D modeling and 3D printing. If you're a hobbyist, engineer, architect, or someone who enjoys building things, there are some great things you will learn from this book. If you have never modeled before, you should start with the very basics in Chapter 1 and work your way through more advanced examples in every chapter. If you're coming in with some experience developing designs using CAD software, then you will find this book a great review. You will learn about developing models and preparing them for 3D printing on Shapeways.

What's Covered

In Part I, you'll take a look at Shapeways, where you will learn how to set up an account, order a model, and see the various materials used for 3D printing. You also are introduced to SketchUp and its tools.

Part II is where all the fun begins. Here you learn how to come up with ideas to model in SketchUp and use SketchUp to design models for 3D printing on Shapeways. You learn what types of models are acceptable for 3D printing.

In Part III, you'll see how to share and sell all the models you have designed in SketchUp throughout this book. You'll also learn how they can be animated and how they can be presented using LayOut. The last chapter concludes the book by exploring other ways you can 3D print your models and you are introduced to Ponoko for laser-cutting your models.

Getting the Most Out of This Book

Learning to model in 3D can be difficult, especially if you haven't had any prior experience sketching or modeling. For a blacksmith, it takes many years of practice to design that perfect tool. Now, I don't expect it will take you years to master SketchUp, but there is a slight learning curve. To assist you along the way, I recommend the following:

- Reading this book is the easy part, but applying what you learn is another thing. So, remember to download Google SketchUp and apply all the examples in this book.

- If you are stuck, refer to the example files for each chapter. You can download example files for this book from the Apress web site. Look at *Beginning Google SketchUp for 3D Printing*'s catalog page in the Book Resources section under the cover image. Click the Source Code link in that section to download the example files.

- Although this book tries to cover all the details you will need to design models for 3D printing in SketchUp, there still may be some things you're probably curious about. If you want to learn more about a particular topic, refer to the appendix, where we take a look at online communities you can join to get your questions answered.

PART 1

Getting Your Feet Wet

■ ■ ■

CHAPTER 1

The Ins and Outs of Google SketchUp and Shapeways

Before you jump in and start the 3D printing process, it's important you learn about the tools used in the development process. With the numerous modeling programs available today, you might ask, "Why should I use Google SketchUp?" I recommend Google SketchUp for two reasons. First, it has an easy learning curve that allows even beginners to create models effortlessly. Second, for such a simple tool, SketchUp offers the user the ability to draw highly complex models. Software such as AutoCAD, SolidWorks, and Blender are also great, but the downside of those modeling packages is the steep learning curve. A beginner might find those software programs intimidating compared to the ease of using Google SketchUp. In the first half of this chapter, I will go through the steps of installing Google SketchUp and introduce you to the SketchUp toolbars.

The second half of the chapter explores the online 3D printing service Shapeways. Shapeways is an online platform for uploading models for 3D printing. By using Shapeways, you avoid having to purchase expensive equipment. The equipment is operated by Shapeways, and all it does is charge you for the cost of the material and service.

Let's not waste any time; by the end of this chapter, the goal is to provide you with enough information to construct a simple model in Google SketchUp and to navigate your way through the maze of options in Shapeways.

Google SketchUp

Google SketchUp is 3D modeling software downloadable from Google. In 1999, @Last Software in Boulder, Colorado, developed SketchUp and then introduced it into the market in 2000, before being acquired by Google. Since then, Google SketchUp has become popular among 3D modelers. What individuals love about this software is its easy learning curve. An individual with no experience in computer 3D modeling could easily learn SketchUp within a few hours of starting the program. With this advantage over other software packages, such as SolidWorks and AutoCAD, individuals with little to no experience can produce designs of immense complexity.

The interface is very simple—a common trait among many of Google's products. This is what especially makes the software package so attractive. The look and feel of the software is what sparked its popularity. It incorporates most of the tools used by software packages such as Microsoft Paint and Photoshop, including Line, Arc, Freehand, Rectangle, Polygon, and 3D Text, which are all common but also powerful tools for modeling 3D structures.

Installing Google SketchUp

You can download either a free or professional version of Google SketchUp online. If you are a beginner or intermediate user of SketchUp, I highly recommend you download the free version, because there is a lot you can do with it without having to pay anything. Most of this book is set up to use the free version of Google SketchUp, but in Chapter 11 we use LayOut, which is a separate software package that comes with the Pro version. I recommend that you wait, read the entire book, and use the free version before investing money in buying the Pro version—unless you are planning to do some serious 3D modeling. The Pro version of Google SketchUp has additional features that allow you to create documents and presentations and comes with an additional tool set called Solid Tools. It also imports and exports a variety of file formats and allows you to create your own custom styles. Now let's get started on our 3D modeling and printing journey.

Downloading Google SketchUp

The free version of Google SketchUp is available for download on the SketchUp home page at http://sketchup.google.com. In the upper-right corner of the page, click Download Google SketchUp, as shown in Figure 1–1.

Figure 1–1. *Google SketchUp home page*

You will be directed to a second page to select between two links (Figure 1–2). You can download Google SketchUp Pro or Google SketchUp 8. For now we will be using the free Google SketchUp 8. Click the Download Google SketchUp 8 link to continue the download process.

Figure 1–2. Google SketchUp free and professional version download page

You are redirected to the agreement page (Figure 1–3). Select the platform you are using, Windows or Mac OS. Then click Agree and Download.

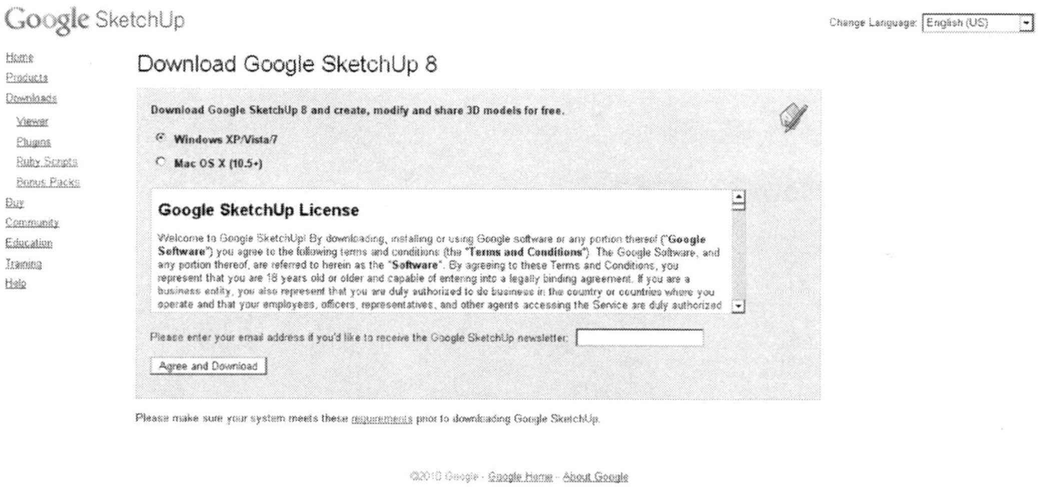

Figure 1–3. Google SketchUp download agreement page

Installing Google SketchUp

For Windows users, a download dialog box will appear; click Save File. Remember to save the file in a location you can easily access. The downloaded file is approximately 40MB, so with a DSL connection, the download will take only a few minutes. After the download completes, double-click the executable file. The Google SketchUp 8 Setup dialog box will appear (Figure 1–4). Click Next to continue the installation.

Figure 1–4. Google SketchUp 8 Setup dialog box

Select the "I accept the terms in the License Agreement" check box, and then click Next (Figure 1–5).

Figure 1–5. End-User License Agreement screen

By default Google SketchUp will be installed automatically into your programs folder. I recommend you stay with the default options at this time unless you have a good idea of where you want to save the file beforehand. Click Next to continue the installation of Google SketchUp (Figure 1–6).

Figure 1–6. Destination Folder screen

Finally, click Install (Figure 1–7). Google SketchUp will take a couple of seconds to install on your computer. After installation is complete, click Finish.

Figure 1–7. Ready to install Google SketchUp 8 screen

User Interface

Once installation is complete, double-click the Google SketchUp icon on your desktop, or select it from your programs menu. When SketchUp opens, you are presented with the Welcome to SketchUp dialog box (Figure 1–8a). On the bottom of the menu, select the Template tab. You are then asked to select a units template (Figure 1–8b). This will be the default template used when drawing models. Since we will be working with small-scale 3D printed models in this book, select the Product Design and Woodworking–Millimeters template. If you do not want the Welcome to SketchUp dialog box to appear again, uncheck the "Always show on startup" box in the bottom-left corner of the window. Then click Start using SketchUp in the bottom-right corner of the dialog box.

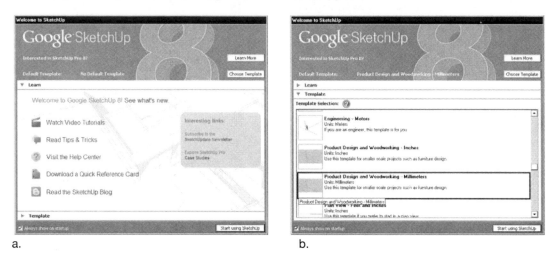

a.   b.

Figure 1–8. (a.) Welcome to SketchUp dialog box; (b.) selecting a units template

When SketchUp opens for the first time after the Welcome to SketchUp menu, the Instructor dialog box will appear. This dialog box gives you an introduction to the Line tool. You will be using the Line tool to construct a lot of the models throughout this book. For now, you can close the dialog box. On first use of SketchUp, you are presented with only the bare minimum needed to start modeling (Figure 1–9). This allows you to begin using SketchUp and avoid being overwhelmed by too many tools, especially if you're a beginner to 3D modeling. As you gradually get comfortable using the basic tool sets, you can easily activate the advanced tools in SketchUp.

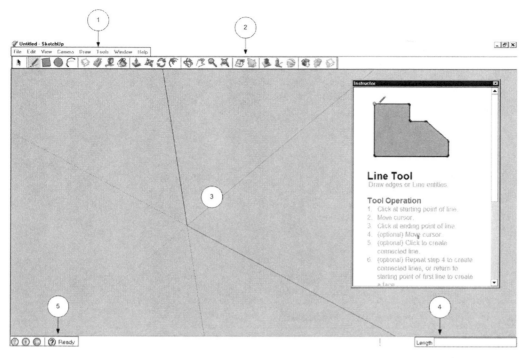

Figure 1–9. *Google SketchUp user interface*

The Google SketchUp user interface is divided into five sections. These are numbered in Figure 1–9 and described in the following list. You will be using each part of the user interface throughout this book, so don't worry about how they work. For now, browse through each menu option and get familiar with their locations in SketchUp. In later sections and chapters, you'll explore each tool's functionality while you develop models for 3D printing.

1. The menu bar is divided into eight menu options. These are File, Edit, View, Camera, Draw, Tools, Window, and Help.

2. The Getting Started toolbar, located below the menu bar, contains the tools you will be using to construct models.

3. The modeling window is where you will spend the majority of your time constructing your 3D models using the tools from the Getting Started toolbar and the Large Toolset. We will be discussing the Large Toolset later in this chapter (see the "Large Toolset" section).

4. The Measurement toolbar is in the bottom-right corner of Google SketchUp. It displays dimensional information about each model and allows you to enter information to adjust model features.

5. The status bar, located on the bottom left of the screen, provides tips on each tool you select while modeling. Keep an eye on the status bar when you are not sure how to use a particular tool while modeling.

Let's continue and learn about the Getting Started tools available in Google SketchUp.

The Getting Started Toolbar

The Getting Started toolbar contains 25 tools, including all the basic tools you will need to start modeling, as shown in Figure 1–10. From the name of the tool, you can probably understand exactly what each does. Table 1–1 explains in more detail each tool's function, based on the numbers in Figure 1–10. In the "Learning by Example" section, you will be applying some of these tools to construct models, and you will get a hands-on feel of how the tools operate.

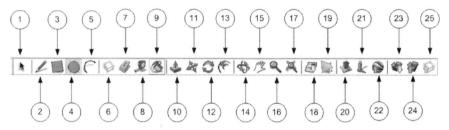

Figure 1–10. Getting Started toolbar

Table 1–1. Getting Started Tools

Tool	Number	Description
Select	1	Selects objects in the modeling window to be modified with other tools
Line	2	Draws lines forming edges and faces
Rectangle	3	Draws a rectangle or square made of four edges
Circle	4	Draws a circle by defining the number of sides
Arc	5	Draws multiple connected line segments
Make Component	6	Creates a component of selected edges and surfaces
Eraser	7	Erases objects from the modeling window
Tape Measure	8	Measures distance, creates guidelines, or resizes models
Paint Bucket	9	Paints objects in the modeling window
Push/Pull	10	Converts 2D surfaces into 3D models
Move/Copy	11	Rotates components and groups and moves and copies objects
Rotate	12	Rotates and stretches objects in a circular path

Tool	Number	Description
Offset	13	Copies lines and surfaces in line with the first model
Orbit	14	Rotates camera around the object
Pan	15	Moves camera horizontally and vertically
Zoom	16	Moves the camera position in or out
Zoom Extents	17	Zooms to the entire model
Add Location	18	Captures site location where the model will appear
Toggle Terrain	19	Toggles between 2D and 3D while in SketchUp
Add New Building	20	Launches the Google Building Maker
Photo Textures	21	Adds photo textures to buildings
Preview Model in Google Earth	22	Places the model in Google Earth
Get Models	23	Downloads models from the 3D Warehouse
Share Model	24	Uploads models to the 3D Warehouse
Upload Component	25	Uploads components to the 3D Warehouse

Large Toolset

When you open Google SketchUp for the first time, you are presented only with the Getting Started toolbar. To activate the Large Toolset, select View ➤ Toolbars ➤ Large Toolset from the menu bar. You will then be presented with 32 additional tools. Hold the cursor over each tool to display its name (Figure 1–11). Some of the tools displayed are similar to those shown in the Getting Started toolbar. Table 1–2 describes the ones that are not present in the Getting Started toolbar, based on the numbers in Figure 1–11.

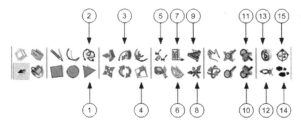

Figure 1–11. *Google SketchUp Large Toolset*

Table 1–2. *Large Toolset*

Tool	Number	Description
Polygon	1	Creates polygonal objects
Freehand	2	Draws hand-drawn lines
Follow Me	3	Extrudes a face along a path
Scale	4	Resizes a model
Dimension	5	Places dimensions in the model
Protractor	6	Measures angles and creates guidelines
Text	7	Adds text to the modeling window
Axes	8	Moves the axes in the modeling window
3D Text	9	Creates 3D text on objects
Previous	10	Goes to previous zoom location
Next	11	Goes to the next zoom location
Position Camera	12	Places the camera at the desired eye height
Look Around	13	Pivots the camera at a single point
Walk	14	Walks within the modeling window
Section Plane	15	Creates cuts to view within your model

The feel of Google SketchUp is very similar to Microsoft Paint. The most common tools you will be using in SketchUp are Move/Copy, Push/Pull, Rotate, Scale, Tape Measure, and Offset. You will be using these tools in most, if not all, of your modeling work throughout the book. With the Move/Copy tool, you can select an object and move it. By pressing the Ctrl key on your keyboard and then selecting the object, you can create a copy of the object. With the Push/Pull tool, you can extrude any surface of the model. The Scale tool resizes the selected surface or the entire model. Use the Tape Measure tool to draw in guidelines to assist you in modeling, or use it to automatically adjust the dimensions of your model. The Offset tool will create a copy of the surface, which you can shrink or enlarge on top of the surface being offset.

We have just gone over some of the common tools you will be using in SketchUp. In the next section, I will demonstrate how to use some of the tools to draw simple objects. I also recommend you take a leap of faith and let your imagination run wild. Select any tool and simply start drawing.

Learning by Example

Let's start modeling by using the tools in the Getting Started toolbar. For our first tutorial, you'll draw simple shapes in the modeling window, which you will then convert into 3D shapes using the Push/Pull tool.

Modeling a Cube

Open SketchUp, and in the Getting Started toolbar, select the Rectangle tool. The cursor turns into a pencil.

1. Take the pencil, click the center axes, and drag the mouse outward.

   As you are doing this, take a look at the Measurement toolbar. The dimensions of the model change as you drag the cursor outward while drawing the rectangle. As you drag the cursor outward, you will see the rectangle get bigger.

2. Click your mouse once more to lock the rectangle in place.

   You can also add a specific dimension rather than dragging the cursor and trying to estimate the size of your rectangle. Instead of locking the rectangle in place, type the value **500, 500**, and hit Enter on your keyboard. This will create a 500mm × 500mm square instead of a rectangle (Figure 1–12). You will recall that a few sections back we set the default unit as millimeters. If you want to work with inches, meters, or feet, type in the desired unit after each number (500", 500", 500m, 500m, or 500', 500').

   After drawing the rectangle or square, if you are unable to see the model, select the Zoom Extents tool from the Getting Started toolbar. This will readjust the modeling window so that the entire model is visible. To show the size of the square in Figure 1–12, I added dimensions to the sides. Dimensions are not placed automatically when you draw a model in SketchUp.

3. To place a dimension, select the Dimension tool from the Large Toolset, and select the edge of the square.

4. Drag the cursor outward, and click your mouse once to lock the dimensions in place.

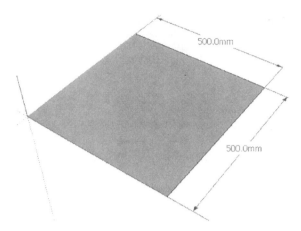

Figure 1–12. Square with the dimensions 500mm × 500mm

The models you draw in SketchUp consist of edges and faces. Edges are made of lines, and faces are the surfaces. The square you just modeled has four edges and one surface.

5. Select the Push/Pull tool, and click the top face of the square. Pull the face in the upward direction, and type the value **500**.

This will create a cube 500mm in height (Figure 1–13). You have just created a box with 6 surfaces and 12 edges.

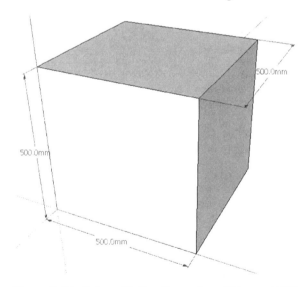

Figure 1–13. Cube with the dimensions 500mm × 500mm × 500mm

I added the dimensions for the cube in Figure 1–13 for better visualization. Just remember to select an edge when adding a dimension to the model.

Creating a Cylinder

Now you will learn how to draw a cylinder. Drawing a cylinder is similar to drawing a cube—the only difference is that a cylinder is made from a circle. The cylinder you will be drawing has a radius of 250mm and a height of 500mm. Take a few minutes, and try to draw one for yourself. Figure 1–14 shows what it should look like after you are finished.

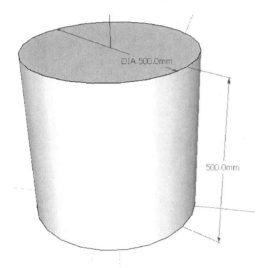

Figure 1–14. Cylinder with a height of 500mm and a radius of 250mm

Here are the steps to draw the cylinder. If you want to save your previous design, please do so at this time.

1. From the menu bar, select Start ➤ New.

    A blank modeling window will appear.

2. Click the Circle tool, and then click the axes.

3. Drag your cursor outward, release the mouse button, type **250**, and hit Enter on your keyboard.

4. Now select the Push/Pull tool, and select the top surface of the circle. Drag the circle upward, release the mouse button, type the value **500**, and hit Enter.

    To move the model to the left or right, you will first need to highlight it.

5. Select the Select tool. Starting from one end of the model, click and drag the cursor to highlight the entire model.

6. Select the Move/Copy tool, click a surface or edge of the model, and then drag the cursor in any direction you desire. Click once more to release the model.

The Rotate tool allows you to orbit around the object for viewing the model from all angles. In addition, it allows you to rotate around the model for editing.

7. Select the Orbit tool, and then click and hold the cursor.

As you move the cursor, the object will also rotate. Release the cursor to stop rotating.

Creating a 3D Pentagon

For the next design, you will be drawing a pentagon. How do you do this? Remember that an object in SketchUp consists of edges and faces. You could draw the entire pentagon using the Line tool, but that would be difficult. There is a built-in function within SketchUp that will automatically draw a pentagon for you.

1. Select the Circle tool, and click the center of the axis.

2. Type **5s**, and hit Enter on your keyboard.

The circle will turn into a pentagon. The "5s" you typed sets the number of sides to 5.

The next time you select the Circle tool, it will still draw a pentagon. To draw a circle, simply type **24s** to increase the number of sides.

3. Type **250**, and hit Enter on your keyboard.

4. Using the Push/Pull tool, click the top surface of the pentagon.

5. Type the value **500**, and hit Enter on your keyboard.

You have just created a 3D pentagon (Figure 1–15).

Figure 1–15. 3D model of a pentagon

Using the Scale Tool

You have just created a 3D pentagon in SketchUp utilizing the Circle and Push/Pull tools. You'll now see how to scale the 3D pentagon. The Scale tool is part of the Large Toolset. As mentioned earlier, you can activate the Large Toolset by selecting View ➤ Toolbars ➤ Large Toolset. The Scale tool is great for resizing models.

1. Highlight the entire pentagon, and then select the Scale tool.

   The object should now be surrounded by a large yellow box with small green boxes (Figure 1–16).

2. Click one of the green boxes, and then drag it in order to scale the model.

   You can also scale the model automatically. For example, type **.5**, and hit Enter to shrink the model by 50 percent.

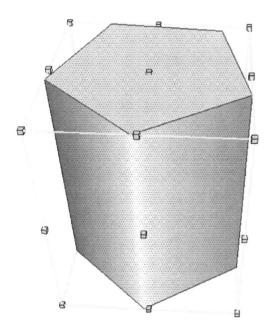

Figure 1–16. *Scaling a model in SketchUp*

Tape Measure Tool

Another great tool within the Large Toolset is the Tape Measure tool. With the Tape Measure tool, you can resize the model with a specific measurement or add guidelines while modeling. For example, Figure 1–17 shows a 20mm radius circle.

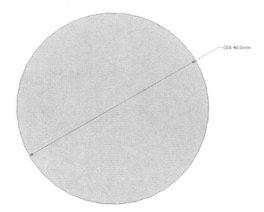

Figure 1–17. Circle with 20mm radius

1. Using the Tape Measure tool, click the edge of the circle, and then click the opposite end of the circle.

2. Type **2**, and then press Enter. The resize dialog box will appear (Figure 1–18).

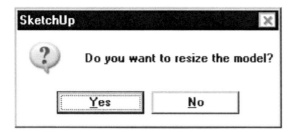

Figure 1–18. Resize dialog box

3. Click Yes. The circle will then shrink to 2mm in diameter.

Remember although this tool will resize the circle to a diameter of 2mm, it will also resize all the other models within the modeling window even though they are not attached to the circle. This tool is useful when scaling your models for 3D printing.

Creating an Offset

The Offset tool is useful when creating hollow objects. As you go through this book, you will notice that you will be creating offsets of all types of surfaces. With the Offset tool, you don't have to draw the surfaces twice. To see how to use the Offset tool, follow these steps:

1. Draw a cylinder; you can draw it to any size you like.

2. Select the Offset tool.

3. Select the edge of the cylinder, and drag the cursor inward.

4. At the desired spot, click to lock the offset in place.

You just created an offset (Figure 1–19a). You can now extrude the surface using the Push/Pull tool (Figure 1–19b).

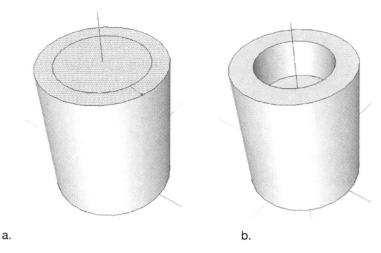

a.   b.

Figure 1–19. (a.) Creating an offset; (b.) extruding the offset

You could have drawn another circle on top of the cylinder instead of using the Offset tool; that's easy when it's just a circle. But if this were an odd-shaped surface, then the Offset tool would be much faster to use.

So far, we have covered some of the basics of Google SketchUp. As we build upon each chapter, I will be introducing additional tools you can utilize and also show techniques you can use to save time when modeling. In the next section, you will take a look at Shapeways, the online platform for uploading and 3D printing your models.

Shapeways

What makes Shapeways so great is that it is one of the first and only online 3D printing platforms available, allowing you to upload models for 3D printing. This means you don't need to purchase expensive 3D printing machinery. The Shapeways web site (Figure 1–20) houses a repository of 3D models for 3D printing. You will find an assortment of models with products in the following categories: Art, Gadgets, Games, Home Décor, Jewelry, Hobby, and Seasonal. The web site is currently in beta and has shown tremendous growth and interest among the design community since 2007. The cost of 3D printing the models on Shapeways is very reasonable. Prices range from a few dollars to a few hundred dollars, but a minimum order of $25 is required. The time from when you order your model to the time it arrives at your doorstep can range from 10 to 21 days depending on the material you choose to 3D print with. You will be exploring the different materials available for 3D printing in the next chapter.

Figure 1–20. Shapeways web site

To access Shapeways, go to www.shapeways.com. You will be directed to the Shapeways web site (Figure 1–20).

Shapeways is not the only site on which you can upload models. There are other web sites similar in many ways to Shapeways, but they do not exactly offer the designer the ability to 3D print models. Ponoko specializes in laser-cutting parts. Thingiverse.com is an online platform for uploading 3D models, but then you need your own 3D printer.

Home Page

If this is your first time accessing Shapeways, you'll notice lots of links, tools, and information that all might seem confusing. Don't worry—I will be showing you how to use each feature on the web site as we go through the book. For now, I will briefly go through all the features presented on the home page and show you how to set up your account on Shapeways.

Menu Options

On the top of the home page are a set of menu options (Figure 1–21). Place your cursor over any one of these, and a drop-down menu will appear with a list of suboptions. On the top menu bar there are five links: home, gallery, community, support, and my designs. Each link is divided into subcategories. For now, hover your cursor over each link and explore the many options. I'll be explaining each in more detail throughout the book.

Figure 1–21. Shapeways' top menu bar

Helpful Links

On the left of the home page are links to information about how Shapeways works, namely, steps for creating buying, selling, and uploading a model (Figure 1–22). Within the Product categories box, you can view a collection of 3D models designers have uploaded.

About Shapeways

How it works

Simple creating

Buy and sell

Upload and 3D print

Product categories	
Art	▷
Gadgets	▷
Games	▷
Home decor	▷
Jewelry	▷
Hobby	▷
Seasonal	▷

Figure 1–22. Left-side menu bar on Shapeways' web site

If you are looking for a particular type of model, use the search bar to type in the keywords describing the model. Shapeways will go through its database of models/parts to find models with similar keywords in the title and description of each model. If you are looking to narrow your search even further, then click the Advanced Search link to the right of the search bar. The "Advanced search" page will appear (Figure 1–23). The form asks for the title, description, tags, category, owner, minimal rating, price, date uploaded, and date last updated. The more parameters you fill in, the more you can narrow your search even further. If you filled in too many parameters and the search comes up blank, then think about taking out some of the parameters entered. If you want to go back to the home page, click the Back button on your browser. Or click the Shapeways logo in the upper left of the page.

Figure 1–23. Shapeways "Advanced search" page

Finding Popular Products

On the home page, if you scroll down, you will find the "Featured shop products" section (Figure 1–24). Here you will find some of the more popular products that users are 3D printing.

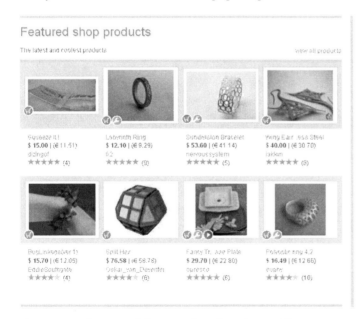

Figure 1–24. "Featured shop products" section on Shapeways

If you zoom in close to one of the featured products, notice the pink and orange circles. The pink circle has an image of a shopping cart, and the orange circle has an image of a spanner. The pink circle indicates the model is for sale. The orange circle allows you to personalize the model. There is also a third black circle—do you see one? This circle indicates there is a video that you view of the model. Watching a video is a great way to see whether the model is worth purchasing. Select any of the images, and you will be directed to the product information page. Here you can get a close-up look of the model and order one for yourself.

Creating an Account

Before you can upload, print, or sell your models, you will need to create a login account on Shapeways. Creating an account is not difficult at all. I'll be guiding you through the process in the next couple of pages. Make note of the account information because you will be using it in the next chapter.

On the upper-right side of the Shapeways home page is the sign-up link. Select the link to be directed to the "Sign up" page (Figure 1–25).

Figure 1–25. Shapeways "Sign up" page

Fill in the sign-up information with a user name, password, and e-mail. Throughout the book you will be logging in and out on a continuous basis while uploading and ordering models, so make sure to choose a user name and password that you can remember. Agree to the terms and conditions and privacy statement. Then click Create Account. If your account is set up correctly, within a few minutes you will get an e-mail to activate your account. Log in to your e-mail account, and activate the Shapeways account by selecting the activation link within the e-mail. Once activation is successful, you can log in to your account with your new user name and password.

Once logged in, the first thing you will need to do is set up your user profile. From the home page, select "my designs," and then from the drop-down list select "My profile." The profile page is divided into three sections (Figure 1–26). You do not have to fill in this information right now, but the billing and shipping addresses are something you will need when ordering models. So, go ahead and fill in the billing and shipping address information. By Chapter 9, you will have a collection of 3D models, and at that time I will show you how to open a Shapeways Shop to sell your creations.

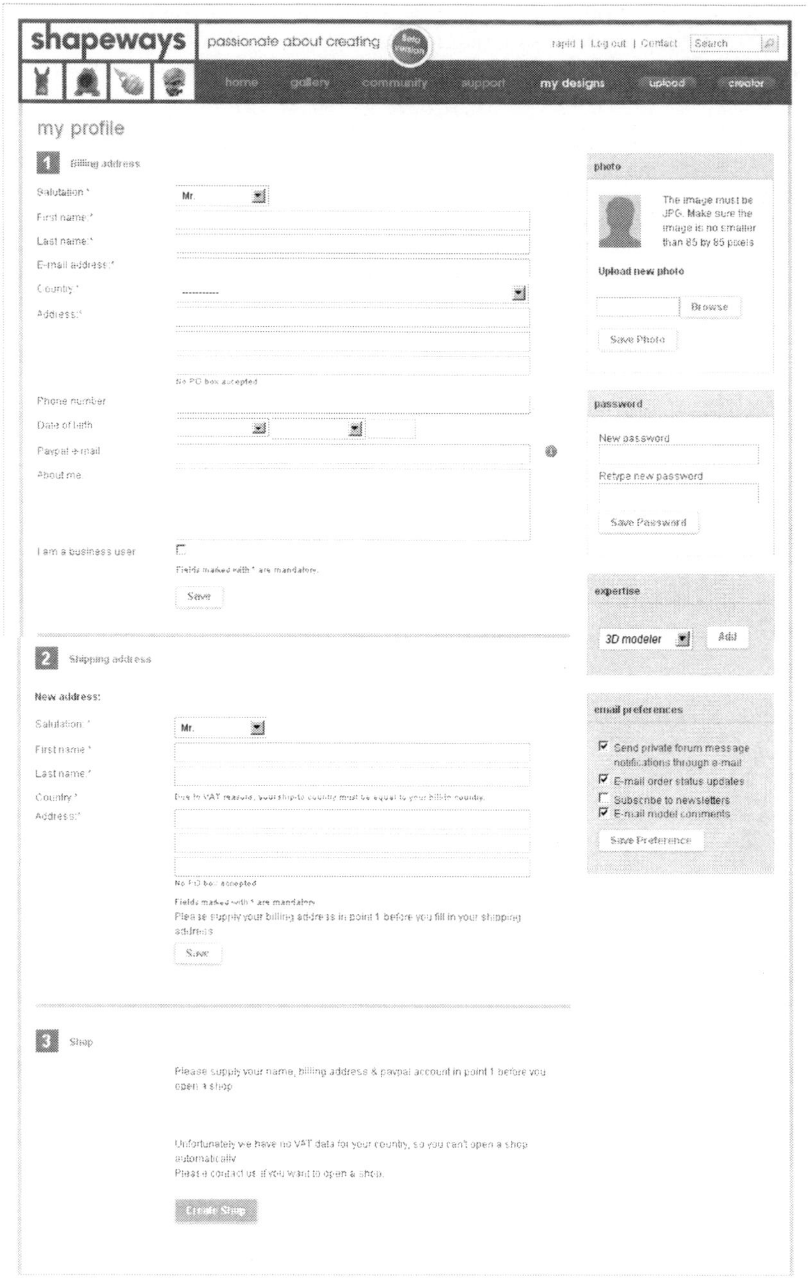

Figure 1–26. Shapeways profile page

On the right of the profile page are a few other options that you can set but are not required. If you are planning on having an active presence on the web site, I recommend you take the time and add those few extra options. For example, you can upload an image of yourself or create a logo. This image will be presented alongside the models you will be posting on Shapeways. I recommend you at least create a logo to upload. People will associate your products with your logo, and it's a great way to market your products. Just below the photo option, you can change your password and select your expertise. In the world of 3D modeling and printing, you will find all kinds of people developing models. There are artists, designers, architects, engineers, and product designers. Choose the expertise that describes you best. Below that are the e-mail preferences. I would stay with the defaults. You will be receiving comments, order updates, and private notifications through e-mail.

Summary

In this chapter, you learned how to download and install Google SketchUp. You learned about some of the different tools and toolbars available for 3D modeling. You also looked at Shapeways and explored some of the links and resources. The chapter concluded by explaining how to create a Shapeways account. In the next chapter, you will be 3D printing your first model using the Shapeways Creator and Co-Creator.

CHAPTER 2

■■■

First 3D Printout

I didn't want to keep you waiting for too long, so in this chapter you'll print your first 3D model using the Shapeways Creator and Co-Creator. "What? I thought this book was about using SketchUp to develop models for 3D printing!" Well, you aren't done with SketchUp yet. You'll be learning how to develop custom models using SketchUp starting in Chapter 4. But while you are reading and getting comfortable with SketchUp, I'll go through the steps and show you how to order models. Shapeways offers a huge repository of 3D models designed and uploaded by other users that you can print. Or you can use the Shapeways Creator and Co-Creator to customize models in just a few minutes. However, if you're excited to learn how to develop models from scratch in SketchUp for upload into Shapeways, then you are welcome to skip ahead to Chapter 4. Finally, we conclude the chapter with a discussion on choosing different materials for 3D printing.

But before you begin, note that Shapeways requires a minimum $25 purchase to complete a 3D print order. Costs will also depend on the material you choose. Refer to the sections "Understanding Model Pricing" and "Selecting the Appropriate Material" for more details.

Getting Started

Open your browser, and go to the Shapeways home page at www.shapeways.com. On the top panel of the Shapeways home page is a set of menu options (Figure 2–1).

Figure 2–1. Shapeways home page menu bar

Select the "creator" button on the far right to be directed to the Shapeways Creator page, shown in Figure 2–2. You can then select from eight different models to customize: Full Color Photoshaper, Photoshaper, Custom Cufflinks, Stylus, Ringpoem, Lightpoem, Fruitconfessions, and Bronze Ringpoem. If a model that you have customized is not to your liking, simply delete it and start again. There are no costs to using the Shapeways Creator. Costs apply only when 3D printing the model.

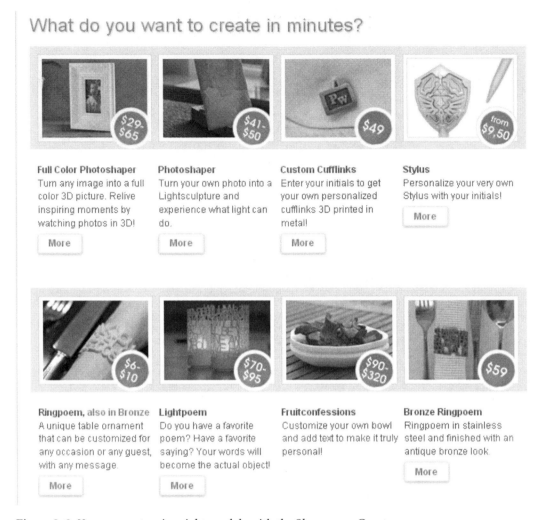

What do you want to create in minutes?

Full Color Photoshaper
Turn any image into a full color 3D picture. Relive inspiring moments by watching photos in 3D!

More

Photoshaper
Turn your own photo into a Lightsculpture and experience what light can do.

More

Custom Cufflinks
Enter your initials to get your own personalized cufflinks 3D printed in metal!

More

Stylus
Personalize your very own Stylus with your initials!

More

Ringpoem, also in Bronze
A unique table ornament that can be customized for any occasion or any guest, with any message.

More

Lightpoem
Do you have a favorite poem? Have a favorite saying? Your words will become the actual object!

More

Fruitconfessions
Customize your own bowl and add text to make it truly personal!

Bronze Ringpoem
Ringpoem in stainless steel and finished with an antique bronze look.

More

Figure 2–2. You can customize eight models with the Shapeways Creator.

The cost of printing a model in Shapeways can be very expensive. The cheapest item you can customize is the Ringpoem, which costs $6–$10. The most expensive item, Fruitconfessions, costs $90–$320. You might be shocked to see that an item so small costs so much. I too was surprised when I first saw the cost of the items. For online 3D printing, this is actually reasonable, though. As the demand for online 3D printed products increases and as new 3D printing technology emerges, the costs of these products are expected to drop. For the sake of cost, we will use the Shapeways Creator to design a custom Ringpoem. The price is very reasonable, only $6–$10. If you are more interested in customizing any of the other products, the steps are pretty much the same as for 3D printing the Ringpoem.

Using Shapeways Creator to Customize the Ringpoem

To get started with the Ringpoem, click More under the Ringpoem image shown in Figure 2–2. You are then redirected to the creator page of the Ringpoem (Figure 2–3).

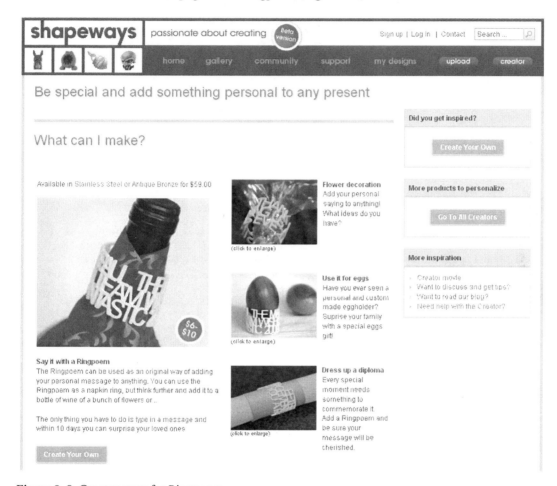

Figure 2–3. *Creator page for Ringpoem*

The page describes ways in which you can use the Ringpoem. You can use the Ringpoem with napkins, bottle of wine, flower, or eggs. To customize the Ringpoem model in the Shapeways Creator, you will need to have an account. Do you remember the user name and password you created in Chapter 1? You will need that information to log in to the Shapeways Creator. If you have not created an account in Shapeways yet, Chapter 1 describes the details of setting up an account in Shapeways.

If you're ready to get started, click the Create Your Own link to be directed to the Shapeways Creator. If you're not already logged in, it will request your user name and password. Type in the user name and password you created in Chapter 1 (Figure 2–4).

Figure 2–4. Sign-in page before accessing the Shapeways Creator

Once you're logged in, a dialog box will appear (Figure 2–5). (The appearance of the dialog box might look different depending on whether you are using Internet Explorer, Chrome, or Firefox.) Click OK to open the Java application on which you will be customizing your model.

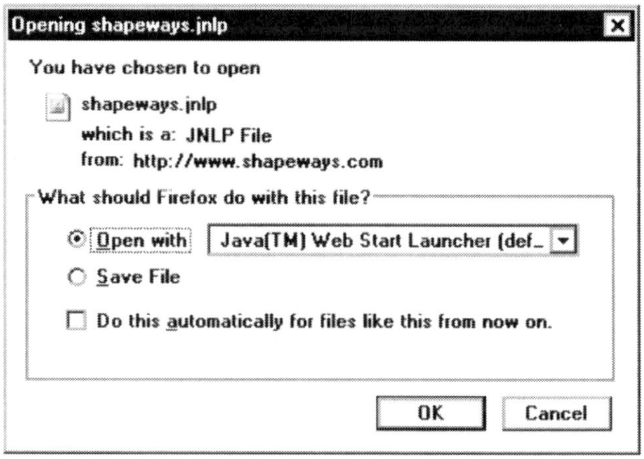

Figure 2–5. The Opening shapeways.jnlp dialog box

The Java application will load, and if everything works as planned, the Shapeways Creator will appear. If the application does not load and an error message appears (Figure 2–6), there could be one of three issues that need to be solved. Either you will need to update the Java plug-in on your computer, clear the Java cache, install Open GL drivers, or do all of the above.

Figure 2–6. Application error message

First make sure you have the Java plug-in installed and updated. To download a copy of the latest version of Java, visit `www.java.com/en/download/manual.jsp`. Select and install the Java plug-in. Remember to download the Java plug-in for your specific operating system. Relaunch the application. Does the same problem arise? If it does, then try to clear the Java cache.

From the Start menu, select Control Panel. Then double-click Java. The Java control panel will appear (Figure 2–7).

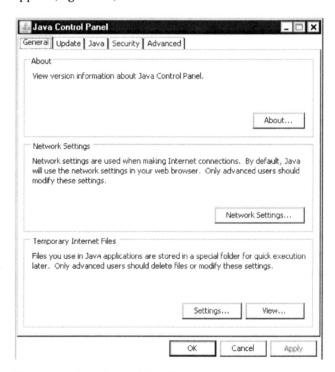

Figure 2–7. Java Control Panel

Click Settings on the General tab. The Temporary File Settings dialog box will appear (Figure 2–8, left).

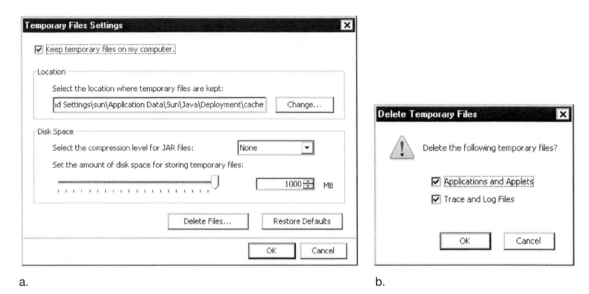

a.   b.

Figure 2–8. Temporary Files Settings and Delete Temporary Files dialog boxes

Click Delete Files. The Delete Temporary Files dialog box will appear (Figure 2–8a); then click OK.

■ **Note** If these techniques don't work, then you might have to update your video card driver. This should be a rare case, unless you are using an old computer. If you do encounter this problem, visit the Shapeways Startup Creator Issues page (Figure 2–9) at www.shapeways.com/support/java_problems. Also search the creator form for possible solutions to the problem you have encountered.

Figure 2–9. *Shapeways Creator startup issues page*

If everything works correctly, the Shapeways Creator will appear (Figure 2–10).

Figure 2–10. *Shapeways Creator application*

The Creator application is divided into four sections: the navigation window, write your message, choose material, and save/order. To customize the model, select "write your message." Within the text box, type the words that you want to appear on the Ringpoem. After typing the message, you might encounter a warning message stating that the text is not structurally stable (Figure 2–11).

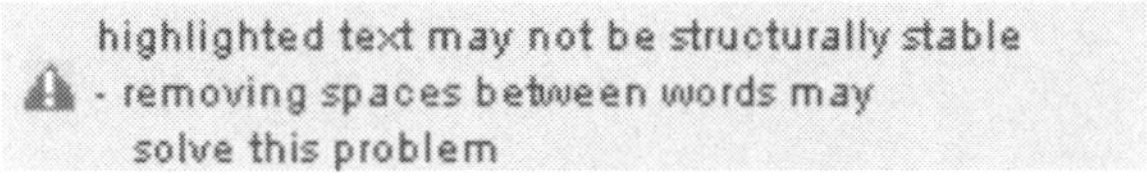

highlighted text may not be structurally stable
⚠ - removing spaces between words may
   solve this problem

Figure 2–11. Warning message

Try writing the message in different forms and selecting a different font type. Usually with a few tweaks, you will have a model that is 3D printable. Then choose a font. Currently there are Tahoma, Trebuchet MS, and Verdana. Now select "choose material." Choose from five different materials: White Strong & Flexible, Black Strong & Flexible, Summer Blue Strong & Flexible, Summer Green Strong & Flexible, and Summer Magenta Strong & Flexible.

At this stage, take a moment to review your design. Now is a good time to adjust the font and material type to reduce the cost of the model. The cost of the model is automatically calculated at the top. Select the arrow keys in the navigation window to rotate the model to view it from different angles. As you rotate the model, make sure to look at all the characters before you save and order the model. The Shapeways Creator is not perfect, and one of the problems you might encounter is overlapping characters. In the model in Figure 2–12, the word *STRUGGLE* has overlapping *G*s. It's very important that you check and then double-check your model for problems. Mistakes can be very costly.

Figure 2–12. The overlapping G *in the Ringpoem*

Saving and Ordering Models

After your final adjustments, you are ready to save and order your model. Click "save/order" within the Creator dialog box. A drop-down menu will appear with three options: save your model, add to cart, and order your model. The first thing you will need to do is save the model before you can order. Click save, and the Shapeways Model Saver dialog box will appear. Save the model with the default name **Ring Poem** (Figure 2–13).

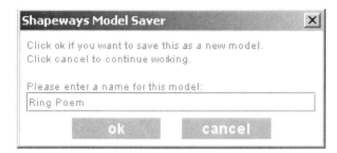

Figure 2–13. Shapeways Model Saver dialog box

Then click "ok." The model will upload to the Shapeways server and be placed in your Shapeways personal gallery for viewing (Figure 2–14).

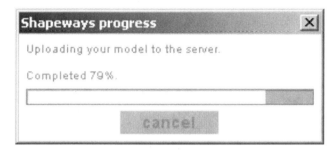

Figure 2–14. Uploading the model to the server

Once the model is saved, an e-mail notification is sent letting you know that the upload was successful (Figure 2–15). If there are any errors in the uploading process, then the e-mail will state the reason for the error. It is a great resource for debugging your model.

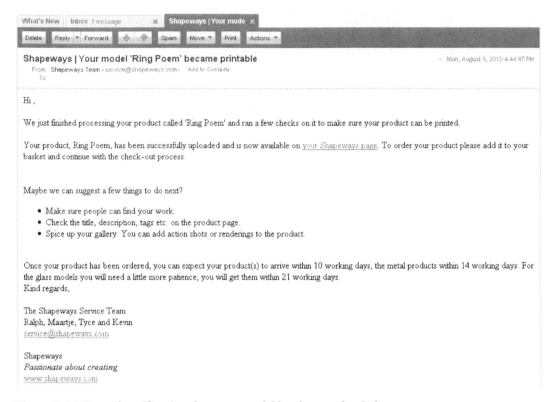

Figure 2–15. E-mail notification that your model has been uploaded

Ordering a Copy of Your Model

You can now order a copy of the model. Within the Shapeways Creator, select "order," and you are sent to the payment page (Figure 2–16). If you have already input your address and billing information as described in Chapter 1, you are ahead of the game. If not, go ahead and fill out the payment information.

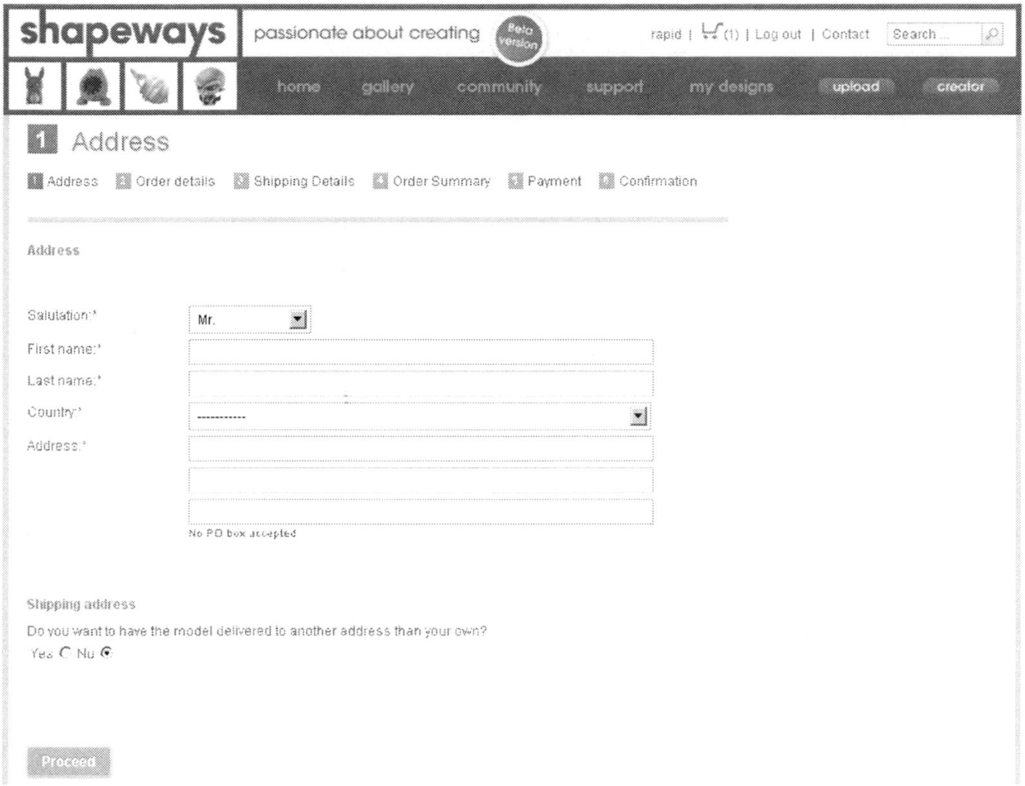

Figure 2–16. Billing information page

After entering the information, click Proceed. You are then directed to the "Order details" page (Figure 2–17). The minimum amount for Shapeways to process the order is $25. The model that you just created costs only $6.09. Before you can precede, you need to order an additional $18.91 of models. Wow! Do you really have to order $18.91 of models? No, it's not required that you print any of the models in this chapter. You can always store the models in your Shapeways gallery and print them at another time once you have a good collection of models you want to print.

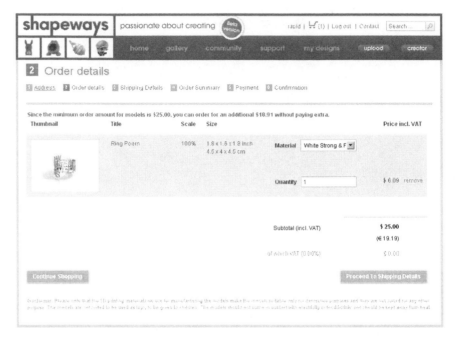

Figure 2–17. Shapeways "Order details" page

Instead of ordering multiple copies of the same product, let's look at some other models from the Shapeways gallery that you can order. You can also customize a product with the assistant of the Co-Creator. Do you remember the following orange circle with the spanner?

Models with the spanner symbol can be customized with the Co-Creator. We just used the Creator, which is a tool developed by Shapeways to customize eight different models on the Shapeways Creator page (Figure 2–2). The Co-Creator, on the other hand, lets you customize models developed by other modelers on Shapeways. In the next section, you will learn how to personalize a model developed by a Co-Creator.

Using Co-Creator to Personalize a Model

The Co-Creator is a way to team up with a designer to assist in the development of your model. It's similar to the Creator, except there is a middleman, the designer, who takes your specifications and modifies the model. The process of personalizing a model involves four steps:

1. Using the Co-Creator platform, choose a design to modify.

2. Enter the specifications of the changes you require in the model; these are changes such as text, images, and measurements.

3. The designer makes those adjustments based on your specifications.

4. Once the adjustments are complete, the designer re-uploads the model to your gallery, and the design is sent automatically for 3D printing.

Let's go through the steps and order a model using the Co-Creator platform. To access the Co-Creator platform, in the top menu bar click "gallery," and from the drop-down menu select "Personalize your own" (Figure 2–18). You will be directed to a gallery of models that you can personalize. This gallery spans several pages. Browse through the selection of models, and find a model you want to print.

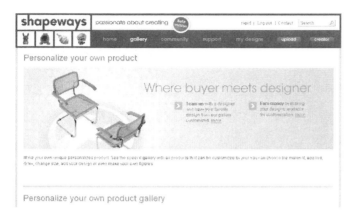

Figure 2–18. Personalize your own product page.

For this example, we will be personalizing the Bracelet V by designer Bulatov (Figure 2–19).

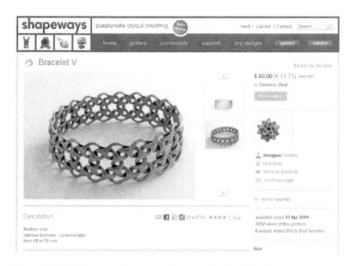

Figure 2–19. Personalize a product with the Co-Creator.

Once you have selected the model and are on the product's gallery page, click the green Personalize link. You will be directed to the personalize page of Bracelet V (Figure 2–20).

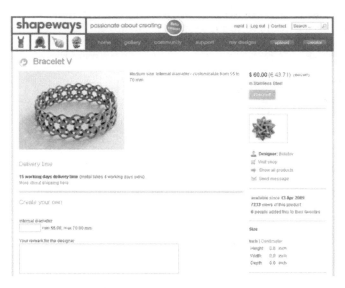

Figure 2–20. Bracelet V personalize page

Personalizing an object is really simple. The only parameter that the designer requires is the internal diameter. After the specifications are input, click Proceed. You can now order or save the model (Figure 2–21). Click Save, and a copy of the model will be saved into your personal gallery for you to order at a later time. Click Order to purchase a copy of the model. You will be directed to the "Order details" page (Figure 2–22).

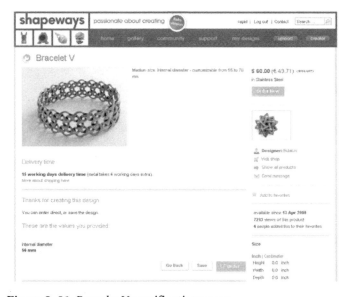

Figure 2–21. Bracelet V specifications page

We are still $1.41 short of the $25 dollar minimum. To meet the $25 minimum quota, you will need to add an extra model to your order. In the next section, you'll add a model from the design gallery available on Shapeways.

Figure 2–22. Shapeways checkout basket

Adding a Model from the Design Gallery

So far in this chapter you have added to your cart models from the Shapeways Creator and Co-Creator. But those are not the only models you can order. Almost all the other models you have seen within product categories are also 3D printable unless they say "Not for sale." To add one of these items, click "gallery" in the top menu bar. From the drop-down menu, select Products. Models in Shapeways can be very expensive. When you select an item to order, think about the cost of investment. In Figure 2–23, I selected the Tiny Hilbert cube variation for $2.73 by the designer Srjskam to add to my order.

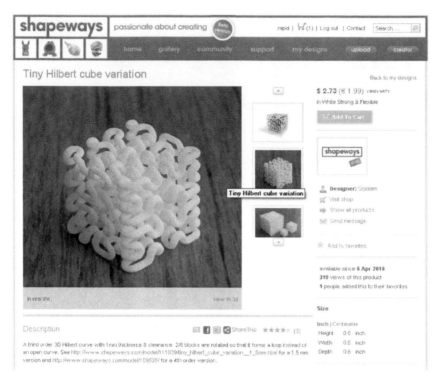

Figure 2–23. Tiny Hilbert cube variation page

Looking at the "Order details" page, the total cost for all three products is $26.32 (Figure 2–24). That's great—just $1.32 more than the minimum price of an order.

Remember to double-check the order. At this stage, if you want to increase the order, change the quantity. Also, this is a great time to change the material you are using. Depending on the material you use, there might be an increase or decrease in cost. For details on the cost of material, refer to the next section. Shapeways will also automatically calculate the cost of the model based on the material you choose before you even order. So, play with the different options and see what works best for you. After you have made the desired changes, click Proceed To Shopping Details. Select whether you want your model gift wrapped. The gift wrap costs an additional $4. Choose either Lovers red or Surprising Blue (Figure 2–25). Then click Proceed To Order Summary.

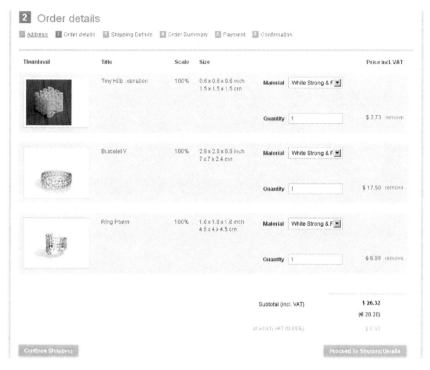

Figure 2–24. *"Order details" page*

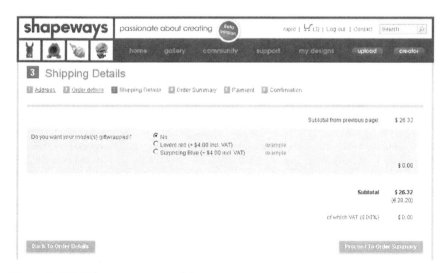

Figure 2–25. *Gift wrap your model*

The order summary page will then appear. Add or change the shipping address information. Double-check your invoice once more before continuing. Then scroll down to the bottom of the page, select your payment options, and agree to the Terms and Conditions (Figure 2–26). Currently Shapeways only accepts PayPal and bank transfers. If you have a credit card, then choose PayPal. You will need to create a PayPal account to transfer money.

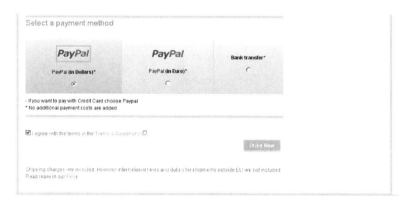

Figure 2–26. Select a payment method.

Click "Order Now," and you are automatically redirected to the PayPal web site. Enter your PayPal account information, and pay for the transaction. After payment, you are automatically redirected to the Shapeways account. Unlike other online stores you might have purchased from, there are no shipping costs. The shipping is included with the price of the model. That's it—you are all done. Your model will arrive packaged in the mail. For models that are printed using stainless steel or alumide, the delivery of the model might take longer since the development process requires more steps to complete.

Selecting the Appropriate Material

The choice of material is very important when 3D printing a model. It is also very important that you consider the application of the product that you have designed. What role is the model going to play? Are you using the model for artistic purposes? Is the model going to be outside where it rains, in the heat, or in the coolness of the night? Are you going to use the model as a hook to hang something? Is a lot of weight going to sit on top of it? When thinking about these questions, think about the type of material that will best work for that particular application. If the model is for decorative purposes, then you might want to add color to the model. If the model will support a lot of weight, then stainless steel might be a good option. If you require a model that is lightweight but also resistant to heat, you could use alumide. Shapeways offers nine material options from which to choose. These are Stainless Steel, Alumide, Glass, Full Color Sandstone, White, Strong & Flexible, White Detail, Transparent Detail, Black Detail, and Grey Robust (Figure 2–27).

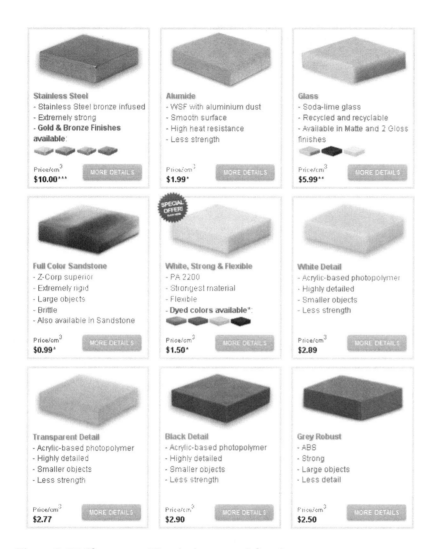

Figure 2–27. Shapeways 3D printing material options

■ **Caution** The material is only for decorative purposes and not to be used for any other purpose. The product should not be used as a toy, given to children, and used as a utensil for eating. I strongly follow their advice and not use the models in any of those circumstances. You will be risking the health and well-being of yourself, your family, and your friends.

Table 2–1 briefly goes over the types of materials available on Shapeways to 3D print your models. The cost of the model you print will depend on the material you choose and the amount you use. When deciding on a material, use Table 2–1 as a reference. All the models printed in this book are using White, Strong & Flexible. A great way to decide what material to print with is to look at what others have been using to print their models. Jewelry is usually printed using stainless steel.

Table 2–1. List of Materials to Choose

Material	Description
Stainless Steel	High strength, available in gold plated and bronze, high heat resistance
Alumide	Medium strength, gray with small speckles, high heat resistance
Glass	Low strength, high gloss, made from soda lime glass, available in white or black
Full Color Sandstone	Low strength, made from gypsum
White, Strong & Flexible	High strength, made from nylon, great heat resistance
White Detail	Medium strength, objects will be smooth, heat resistance greater than 50 degrees Celsius
Transparent Detail	Medium strength, objects will be smooth, heat resistance greater than 50 degrees Celsius
Black Detail	Medium strength, objects will be smooth, heat resistance greater than 50 degrees Celsius
Grey Robust	High strength, objects might have texture, suitable for large objects

Understanding Model Pricing

Several factors go into the cost of each model. Shapeways automatically calculates the price of each model based on each of these factors. When you order a model, it shows only the final cost, as shown earlier in Figure 2–24. There is a startup cost to print each model. The startup cost depends on the material you choose for the model. For Alumide, Full Color Standstone, and White, Strong & Flexible, the startup cost is $1.50. For the Milky White Matte Glass, the startup cost is $5. For the dyed colors such as Blue, Magenta, Green, and Black, the startup costs is $4 and an additional $1.99 per cubic centimeter that your item uses. For Night Black, it is $1.78 per cubic cm. For some of the more expensive items, there are large price jumps. Stainless Steel has a startup cost of $10. Gold Plated and Antique Bronze models are expensive. These models require several steps of production. In each step, the Stainless Steel is infused with bronze and then cured. To produce a gold finish, the models are submerged in a gold bath. The bronze models are oven-baked. At the end, both models are polished (Figure 2–28). Since there are multiple steps to production, expect to wait 18 working days before you receive your product. For specific details on Gold Plated and Antique and Bronze models, visit www.shapeways.com/materials/stainless_steel_finishes.

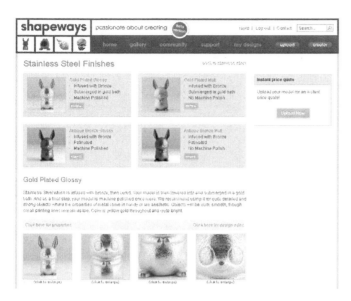

Figure 2–28. Stainless steel finishes

Still not sure what material to print your models out of? No worries—Shapeways provides a sample kit of material (Figure 2–29). The sample kit consists of the following material types: Grey Robust, Transparent Detail, White Detail, Black Detail, Alumide, White, Strong & Flexible, and Stainless Steel. Visit the following link to take a look at the materials kit:
www.shapeways.com/model/133452/shapeways_materials_sample_kit.

Figure 2–29. Shapeways materials sample kit

The sample kit costs $30. Yes, it is expensive, but on the plus side, they do give you a $25 discount coupon for your next order.

Summary

In just a single chapter, we covered a lot of details about Shapeways. You learned how to personalize models using the Shapeways Creator, how to use the Co-Creator, and how to order models on Shapeways. We concluded the chapter with a discussion on model pricing and the types of materials available on Shapeways.

PART 2

Starting from Scratch

CHAPTER 3

■ ■ ■

Getting Your Juices Flowing

You've just seen how to take existing models and prepare them for 3D printing. It was easy enough, because the models were already created. For the first-time designer, though, coming up with ideas for 3D (or even 2D) modeling can be difficult. Even creating a model in Google SketchUp can be a challenge. As a 3D designer, you will come across many problems, especially when developing a single model. That's because many designers are bombarded with an influx of ideas. Instead of assisting, these ideas can end up paralyzing a designer's mind. On the other hand, some designers just cannot think of anything to model. Designing can become a very frustrating task if you do not learn how to handle some of these issues. It is a tedious task and requires a lot of patience. In this chapter, you'll quickly explore some of the ways to overcome these challenges. In doing so, you will be well-equipped to forge ahead with your ideas.

Brainstorming Techniques

When writers have difficulty coming up with new ideas, they call this "writers block." I call the inability to come up with ideas to model "designer's block." Fortunately, as your modeling skills improve, you will need to know about techniques to assist you in generating new ideas. In this section, you will learn some useful brainstorming techniques to apply in your design work to overcome designer's block.

Mind Mapping

A *mind map* is a great way to come up with ideas to model. To create a mind map, take a blank sheet of paper. In the center of the page, write a word—it can be any word that you are thinking of right at this moment. Now in three minutes, create branches and subbranches of similar words. You will be amazed at what you are actually capable of drawing with time constraints. After those three minutes are up, look back at the piece of paper and see whether there is anything of interest. The mind map in Figure 3–1 starts with the word *House* and branches off into several subbranches. Open your mind, and be as creative as possible.

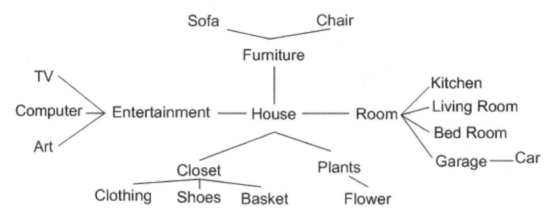

Figure 3–1. Mind map

Gap Filling

If you are designing a large project that has multiple parts, another way to ease the pressure of designing is using a technique called *gap filling*. With gap filling, you first write down a starting goal and an ending goal. Then you list all the ideas that fill in the gap defining those two goals. Figure 3–2 shows gap filling being used with the words *Car* and *Wheel* as the start and end goals. Simply fill in the gap to define the car. A car has thousands of parts. The list could be endless.

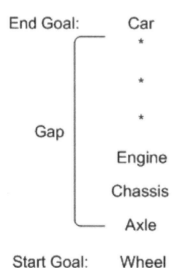

Figure 3–2. Gap filling exercise

Reverse Thinking

Another great approach is using *reverse thinking*. Think of an idea and then simply think about its opposite. To assist you in the process, use a thesaurus. Visit www.thesaurus.com, type in **fan**, and Thesaurus.com will display all the words similar to *fan* (Figure 3–3). If you were thinking of an air conditioner, what could be its opposite? A heater and a furnace are great examples.

Main Entry: fan

Part of Speech: *noun*

Definition: blower of air

Synonyms: air conditioner, blade, draft, flabellum, leaf, palm leaf,
                         propeller, thermantidote, vane, ventilator, **windmill**

Roget's 21st Century Thesaurus, Third Edition
Copyright © 2010 by the Philip Lief Group.
Cite This Source

Figure 3–3. Using the thesaurus to come up with ideas

These are few of the brainstorming techniques you can apply to generate ideas while you are modeling. If you are looking for additional brainstorming sources, an online search of the keywords *brainstorming techniques* will present you with some great ideas.

Taking Advantage of Pencil and Paper

To avoid the frustration of later having to redesign your model as a result of an error you did not catch, let's explore how pencil and paper can save you time and money in avoiding those problems. Many designers come across frustrating moments halfway through their designs, realizing they need to start all over again. They realize the model they just created isn't exactly what they wanted. It's similar to writing the first draft of an English paper. Usually the first draft isn't anywhere close to what you actually wanted it to be. So, you write a second or third draft to refine your thoughts and ideas. Many of these issues could be avoided if we took the same approach. In this section, using pencil and paper, you'll draw multiple sketches of your model before coming up with a final design.

Grab a pencil and multiple sheets of paper, and think of an object you would like to model in SketchUp. On the first sheet of paper, sketch the model that comes to mind. Repeat this process two more times, but each time think of a new way to design the same model. With this approach, you waste little time and are able to refine the design of your model. With some practice, you will start developing a technique that works best for you. So, remember to first draw your model on paper. If you are not happy with your drawing, then draw a few more designs. Compare them, and make edits as needed. Figure 3–4 shows an example of three laptop coolers hand-sketched and used as the basis to later design in SketchUp (Figure 3–5). This is a great way to avoid a lot of design mistakes. I know the task of having to redraw multiple models on paper might seem boring, but trust me, you will be much happier later.

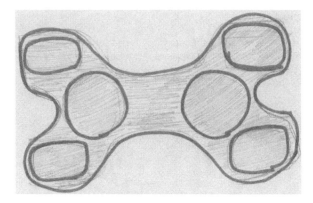

a.

b.

c.

Figure 3–4. Three original sketches of the laptop cooler

From all three concept models in Figure 3–4, I chose sketch (a.) to model in SketchUp after making a few modifications (Figure 3–5). Drawing the model on paper first allows you to quickly and easily fix problems early in your design. Also, it allows you to come up with some great new ideas in the process. Great designs are developed through the process of iteration.

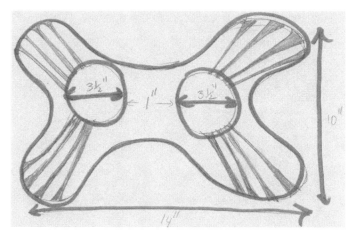

a.

b.

Figure 3–5. Laptop cooler sketched and then modeled in SketchUp

Design Ideas

As a good designer, you will want to share your designs with others. Show your designs to friends, family, and especially someone familiar with CAD design systems and CAD modeling. By sharing your designs, you get feedback and usually insights for improvement. Another great resource for finding ideas is to browse through books at your local library. The next time you're at your local library, grab architectural, arts/crafts, and mechanical books, and search for any great ideas within them that are of interest. There is an abundance of information out there. When coming up with ideas to model, try to keep an open mind. You are not limited to mechanical, arts/crafts, or architectural projects. There are also interesting possibilities for working with molecular models from chemistry, biology, and physiology.

Patents

Another great option is using Google Patent Search at www.google.com/patents (Figure 3–6). You can search more than 7 million patents. Try searching the U.S. Patent and Trademark Office (USPTO) home page at www.uspto.gov. Personally, I have found Google Patent Search more user-friendly, but check out both to see what you can find. In addition, Google has applied optical character recognition (OCR) on each page, so finding what you're searching for is easier.

Figure 3–6. Google Patent Search web page

In Google Patent Search, type the word **ruler**, and select from an assortment of rulers to model. Figure 3–7 shows a ruler modeled in Google SketchUp using a patent. You can search for this patent by directly typing in the patent number. For the model in Figure 3–7, the patent number is D260005. Type it into the patent search engine to see if you can find it.

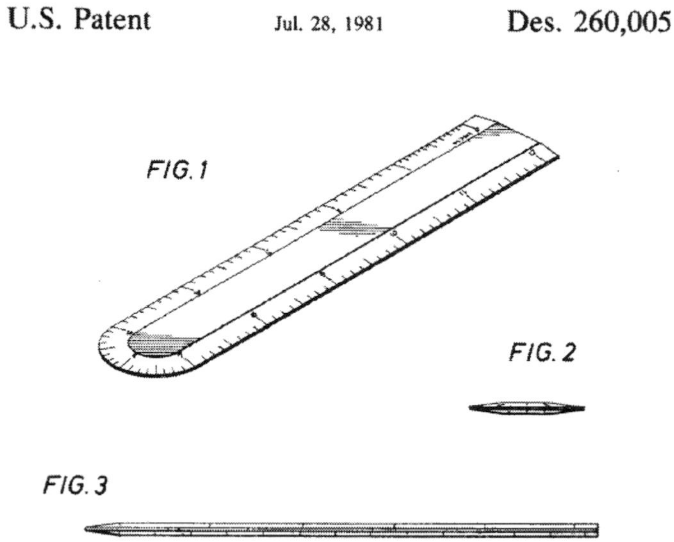

U.S. Patent   Jul. 28, 1981   Des. 260,005

FIG. 1

FIG. 2

FIG. 3

a.

b.

Figure 3–7. Ruler modeled in SketchUp from U.S. patent 260,005

Another great feature in Google Patent Search is the order in which information within each patent is laid out. Each patent page in Google Patent Search is divided into six sections for readability and searchability by users, as listed in Table 3–1.

Table 3–1. Sections Within Google Patent Search

Title	Description
Patent Summary	This is a brief summary of the details of the patent.
Claims	This page states what the patent is protecting.
Drawings	This is an image of the object patented.
Search within this patent	You can search keywords within a patent.
Citations	This page lists other patents that were cited by the given patent.
Referenced by	This page lists patents that are referencing the current patent.

You can find a detailed description of each section and much more on the Google Patent Search Help page at www.google.com/googlepatents/help.

Google 3D Warehouse

Similar to Google Patent Search, there is the Google 3D Warehouse. Rather than house a collection of patents, the Google 3D Warehouse has a large repository of SketchUp models that you can search and download. These models have been designed by other SketchUp modelers. The 3D Warehouse is a great resource for brainstorming models to develop. Visit the Google 3D Warehouse at http://sketchup.google.com/3dwarehouse/, and browse through the collection of models available. I will be discussing Google 3D Warehouse in Chapter 9.

Photographs

If photography is what you really enjoy, then pop out your photo album, or start taking pictures of buildings and landscapes to use in your modeling work. Within photographs, you can capture all the detail you need for your model. If you have done some traveling, you probably have taken many pictures from around the world. It must have inspired you, you can draw on that inspiration to enhance your modeling ideas. In Chapter 7, you'll use the Match Photo feature in Google SketchUp to construct a house using a photograph.

Games

If games are what you enjoy, why not make up a new board game? Off the top of your head, can you think of some cool board games available on the market? There are the classic board games such as chess, checkers, Clue, and Monopoly—but who knows, you might be the one to invent the next big game. You are not just limited to board games either. Think about the many types of pool, ping-pong, and air hockey tables that you might develop designs of in SketchUp. One game that I really enjoyed playing as a kid was Carrom. Carrom is very similar to pool, except everything is smaller. Instead of using a pool stick, you use your hands to flick each Carrom piece into one of the four holes in a corner of the board. On Wikipedia, you can find an assortment of information about the origins, rules, and regulations of the game at www.en.wikipedia.org/wiki/Carrom. Figure 3–8 shows a design of a Carrom board modeled in SketchUp.

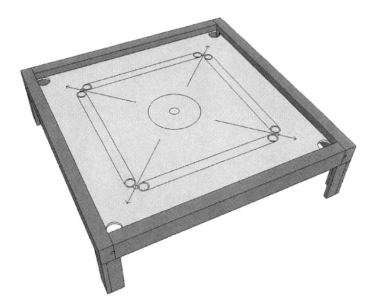

Figure 3–8. Carrom board

Summary

Though this was a short chapter, the goal was to introduce you to the techniques of coming up with ideas for things to model. You learned several brainstorming techniques, how sketches can assist in the modeling process, and how using Google Patent Search, photographs, and games can assist in the idea process. Keeping these brainstorming techniques in mind in the next chapter, you'll create your first model in Google SketchUp for 3D printing.

CHAPTER 4

3D Model to 3D Print

It's time to bring the ideas you've learned in the first couple of chapters and develop a model in SketchUp for 3D printing in Shapeways. The goal of this chapter is to get you acquainted with all the basic steps of developing a model for 3D printing. You'll start this chapter off with a brainstorming session where you develop sketches of the model. Then you'll learn about some of the dos and don'ts for developing models for 3D printing before constructing the model in SketchUp. The chapter will conclude with methods of double-checking the model before sending it off for 3D printing. So, let's get started with our adventure.

Brainstorming a 3D Model

In this chapter, you will be 3D modeling a house. Houses come in different shapes, sizes, and styles. The difficulty arises in deciding what type of house to draw. Let's use a brainstorming technique from Chapter 3 to decide on a model to design—mind mapping. Grab a sheet of paper, and in the middle write the word *house*. Now, in a three-minute time frame, think of different houses you would like to model. Figure 4–1 shows an example of what your mind map could end up looking like.

Figure 4–1. Mind mapping the word house

Now that you are done with your mind map, pick a word that stands out. From the keywords, I have chosen the word *lighthouse* to model in Google SketchUp. Before you continue, take out a pencil and a few sheets of paper. You will be sketching a couple designs of the lighthouse. By sketching the designs, you are able to dump your thoughts onto paper and look at your designs from a different perspective, overall making a better decision. I recommend drawing the sketches on a piece of graph paper. The grid on the graph paper will assist you in creating proportionally sized models. Drawing sketches is not about pulling out your ruler and making sure every line is straight. The goal is to draw models as quickly as possible to come up with several ideas.

Figure 4–2 shows three sketches of the lighthouse. By hand-sketching these models, you can easily edit them. You are able to improve your model through multiple sketches. And lastly, you are able to anticipate many of the problems you may encounter and therefore avoid them during the modeling process. With this process, you save a lot of time and effort.

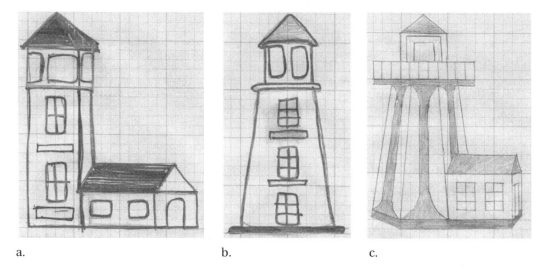

a.   b.   c.

Figure 4–2. Three sketches of the lighthouse

Modeling the Lighthouse

From the three sketches in Figure 4–2, let's say we decide to model sketch b in Google SketchUp. The task of modeling a lighthouse might seem overwhelming at first, but do not worry—I will go through each step of constructing the model with example illustrations. But before you continue, you will need to ask yourself these five questions when developing any 3D model:

- Is the model closed?
- Are the white surfaces facing outward?
- Is the model manifold?
- Does the model meet the specifications for the material?
- Is the model structurally stable?

If you can answer yes to all of these questions, then you are in good shape. You have avoided most of the major problems when developing models for 3D printing. Let's go through a couple of examples, seeing places where these rules would pass or fail.

Dos and Don'ts

All the models developed in Google SketchUp need to be closed, meaning that the models you design should not have any holes in them. For example, a box cannot have an opening on any of its sides. An opening would not define a complete model, and the 3D printer will not fill in the inner parts while printing. You can still make a box with only six sides, but there has to be depth in the model so that the 3D printer knows where to add material.

Is the Model Closed?

Figure 4–3 shows examples of printable, closed models and nonprintable, open models. The two boxes on both ends are printable, but the two center boxes are not. The box on the left is sealed on all sides with no holes. The box on the right is hollow but also has a wall thickness where material can be deposited. At first glance, it looks like there is a hole in the model. But since all the surfaces have white shading, the box is closed. These two boxes pass our first rule. The two boxes in the middle do not have a defined wall thickness and as a result fail the first rule.

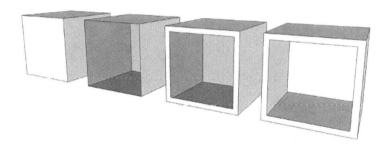

Figure 4–3. Example of printable and nonprintable boxes

Are the White Surfaces Facing Outward?

You know that the white surfaces of the two outer boxes are facing outward since that's all that is visible. In the case of the two inner boxes, you can see a visible darker inner shading. Unfortunately, Shapeways will detect this as an error, and the model will not upload. It looks as if the white faces are pointing out, which they are, so why would it fail uploading? Since the dark shading is visible, there could be one of two problems. The model could have no wall thickness, or one or more walls of the model could be flipped and pointing out. To reverse a surface, right-click its surface, and from the drop-down menu select Reverse Faces.

Is the Model Manifold?

Another common problem you will encounter is called *nonmanifolds*. Nonmanifolds are created when an edge shares more than two surfaces. Figure 4–4 shows an example of a nonmanifold. Can you see the problem? Look at the image on the right. What you see is an extra internal surface. The edge of the circle intersects with three surfaces. Deleting the internal surface will solve the problem.

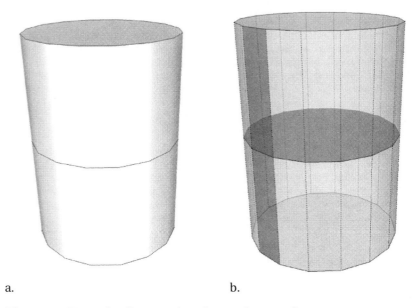

a.   b.

Figure 4–4. *Example of intersection of more than two faces*

Does the Model Meet the Specifications for the Material?

The thickness of your model is also important for 3D printing. To avoid a rupture in the model, you need to consider the maximum model size and the minimum wall thickness (Figure 4–5). For details about the maximum model size and minimum wall thickness, see the Materials Comparison sheet at www.shapeways.com/materials/material-options. Here you will find details on the specifics of each material and their constraints. For the model you will be constructing in this section and throughout this book, you will be using White, Strong & Flexible. The minimum wall thickness of this material is 2mm, and the maximum wall size is 310mm ×230mm×180mm.

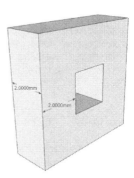

Figure 4–5. *Recommended wall thickness no smaller than 2mm*

Is the Model Structurally Stable?

The overall design of your model is also very important to think about and usually is overlooked by many individuals. If the model has a greater mass on one end than the other, the model might break during the cleaning, packaging, and shipping process. In Figure 4–6, the center post of the model is much smaller compared to the outer walls on both ends. If the walls are too big, the post might break from the stresses applied by the wall. A method of overcoming this problem is printing your model in pieces. Or simply design the model, making sure each part is proportionally sized to meet the stability requirements. In Figure 4–6, the center beam could be enlarged to support the weight of the top surface as well.

Figure 4–6. Center post might break due to uneven distribution

Remember to keep these things in mind when modeling anything for 3D printing. You will be applying these rules in this section and throughout the book as you design different models. Let's get started.

Constructing the Model

Open a new modeling window in Google SketchUp. To keep the modeling costs low, you will need to make sure that the model is not too big. Let's say the model will be no larger than 52 mm (L) × 52 mm (W) × 78 mm (H), or 2.05 in (L) × 2.05 in (W) × 3.07 in (H). In case the model does get too big, you can always scale it smaller.

You will start with the creation of the foundation and build your way up, finishing with the design of the door and windows of your lighthouse in the following order:

- Foundation
- Tapered wall
- Balcony
- Lantern room
- Lantern room windows
- Tapered wall door
- Tapered wall windows

At this time, it also a good idea to get familiar with the Camera toolbar and especially the Orbit, Pan, Zoom, and Zoom Extents tools (Figure 4–7). When maneuvering in SketchUp, these tools will be very helpful. The Orbit tool allows you to easily change the orientation of the camera in the modeling window so you can view all sides of the model. Using the Pan tool, you can move the camera left, right, up, and down. Select Zoom to zoom in and out of the model. Use Zoom Extents to zoom into an entire view of the model.

Figure 4–7. Camera toolbar

Creating the Foundation

From the Getting Started toolbar, select the Circle tool, and click once in the center of the axis. As you drag your cursor outward, a circle will appear. You actually want an octagon as the base for the lighthouse. An octagon is made up of eight sides. Type **8s**, and hit Enter on your keyboard. The circle should turn into an octagon. Type **26mm** and hit Enter again to define the radius of the octagon (Figure 4–8).

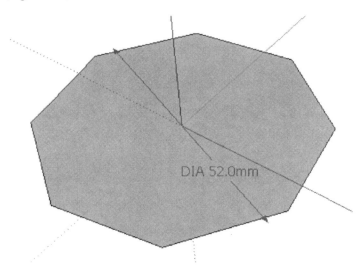

Figure 4–8. Model the base of our lighthouse

You have just created the base for the lighthouse. For ease of modeling, next you will rotate the octagon so that it is in alignment with the red, green, and blue axes. This will be helpful when you construct the door of the lighthouse later in this chapter. The solid blue line points up, and the dashed blue line points down. The solid red line points east, and the dashed line points west. The solid green line points north, and the dashed line points south. Unless you are developing models for Google Earth or casting shadows, you don't need to be concerned about the model's real-world location.

Rotating the Model for Alignment

From the menu bar, select Camera ➤ Standard Views ➤ Top (Figure 4–9a). You should now see the top view of the octagon. Choose the Select tool, and then click and drag to select the entire model. The model should be highlighted in blue. Select the Rotate tool, and click in the center of the octagon; then click once more in the middle of one of the sides of the octagon. You want to select the middle of one of the sides since it will be easier to model the door of the lighthouse. The Rotate tool will lock in place, and now you should rotate your cursor until the protractor is aligned with the axis. Click once more to release (Figure 4–9b).

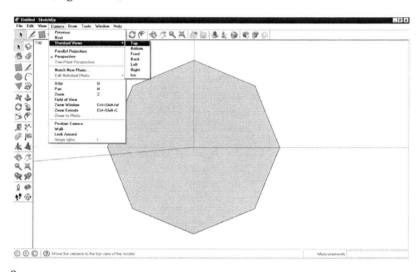

a.

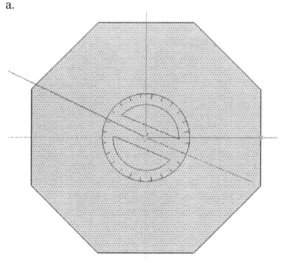

b.

Figure 4–9. View the model from the top, and rotate the octagon with the Rotate tool.

Adding the Tapered Wall

Now rotate the model so that it is in isometric view. From the menu bar, select Camera ➤ Standard Views ➤ Iso. Now you will raise the octagon along the solid blue line at a 2-inch height. From the Getting Started toolbar, select the Push/Pull tool, and click once on the surface of the octagon. Move the cursor along the the solid blue line, and notice the 2D surface is changing into a 3D model. Type **50.8mm**, and hit Enter on your keyboard (Figure 4–10).

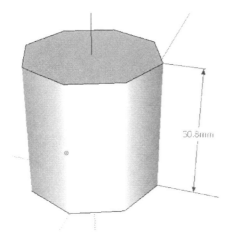

Figure 4–10. Extruding the octagon

Most lighthouse towers taper off at the top. You will need to taper the top of the octagon as well. Select the Scale tool from the Large Toolset, and click the top surface of the octagon. A yellow box with small green squares appear, which is actually normal (Figure 4–11).

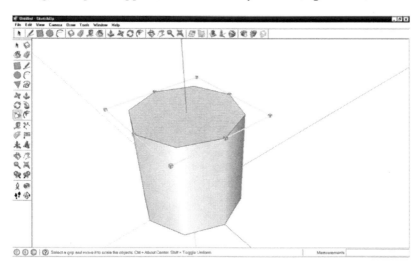

Figure 4–11. Using the Scale tool to taper the model

Place the cursor over one of the green boxes in the corner, and the box to its opposite will turn red. Hold down the Ctrl key on your keyboard, select the green box on the corner, and drag it inward. Drag the box inward until the Measurement toolbar reads .60 (Figure 4–12). Holding down the Ctrl key tapers all the sides of the model at once. Try to taper the model without holding the Ctrl key. Only one side of the model will taper. Select Edit ➤ Undo to go back to the previous step in case you taper only one side.

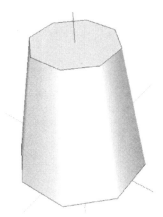

Figure 4–12. Tapered model using the Scale tool

Adding the Balcony

The next step is to extrude the top surface another 2mm. Select the Push/Pull tool, and extrude the top surface by 2mm (Figure 4–13).

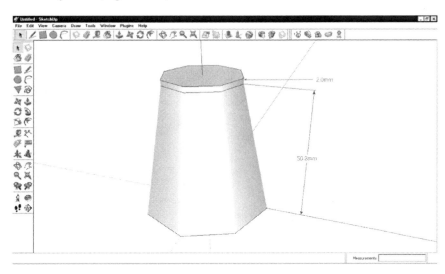

Figure 4–13. Top surface extruded with the Push/Pull tool

The top portion will be the balcony for the lighthouse. Using the Offset tool, you will be creating a 2mm offset of the top surface. Select the Offset tool, and click the top surface. As you drag the cursor to the center of the model, it will create a smaller outline of the surface. Type **2**, and hit Enter. This will create a 2mm offset from the edge of the top surface of the model (Figure 4–14).

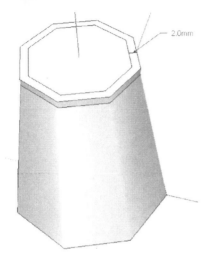

Figure 4–14. *An offset on top of the model*

To create the balcony in your model, you will extrude each of the surfaces on the side. For this next part, you will have to active the hidden lines. The hidden lines define the edge of each surface. In this model, there are multiple surfaces that define the octagon tower. To access these surfaces, you will have to turn on Hidden Geometry. From the menu bar, select View ➤ Hidden Geometry (Figure 4–15).

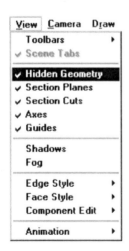

Figure 4–15. *Turning on Hidden Geometry*

Now that the hidden lines are visible, you can use the Push/Pull tool to extrude each surface on the side (Figure 4–16a), defining the balcony by 5mm. The top of the tower should now look like a flower petal (Figure 4–16b).

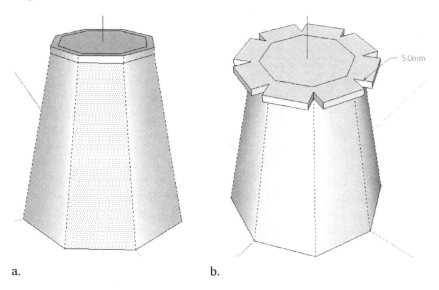

a.   b.

Figure 4–16. Extruding the side surfaces by 5mm

Now you combine each of the pedals. Select the Line tool, and connect the corners of each petal (Figure 4–17a). Repeat this for all the petals. Then delete the lines, creating a single surface (Figure 4–17b).

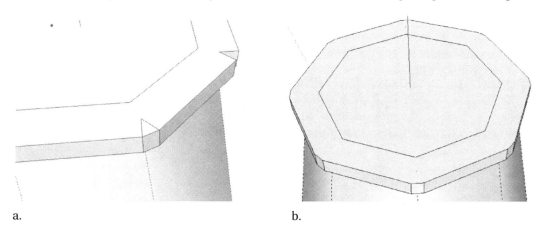

a.   b.

Figure 4–17. Connecting the petals on the model

Adding the Lantern Room

Now that the balcony is complete, the next step is to create the lantern room. First extrude the top surface of the lantern room by .75 inches (Figure 4–18a). Then using the Line tool, create diagonal lines from each corner of the top surface (Figure 4–18b).

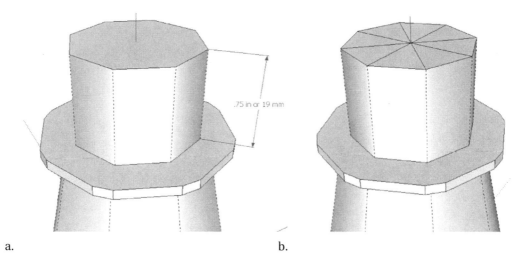

.75 in or 19 mm

a.   b.

Figure 4–18. Extruding the center surface and creating diagonal lines

Select the Move/Copy tool, and click once on the center of the top surface. Move the cursor along the solid blue line, and notice how the top surface changes. Type **.25"** and hit Enter to create the roof of the lighthouse (Figure 4–19).

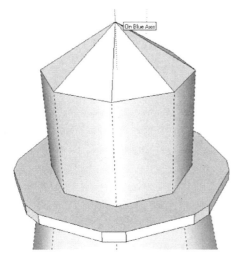

On Blue Axis

Figure 4–19. Extruding the diagonal lines to create the roof

Defining the Lantern Room Window Area

The next step is to create the boxes that define the window of the lantern room. Once again, select the Offset tool, and create a 1mm offset of each surface defining the lighthouse window (Figure 4–20a). Then using the Push/Pull tool, create a 1mm inward extrude of each offset (Figure 4–20b).

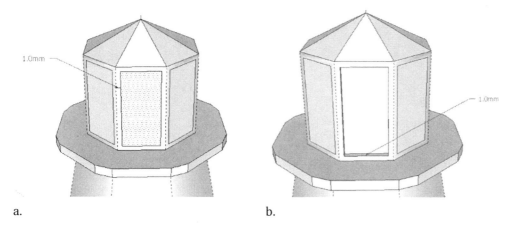

a.   b.

Figure 4–20. Lighthouse windows created with the Offset and Push/Pull tools

Creating the Tapered Wall Door

Next we will create the door for our model. First draw two lines 20mm each that are in parallel with the dashed lines starting from the base of the model. Connect the two lines, forming a box on the side of the lighthouse (Figure 4–21).

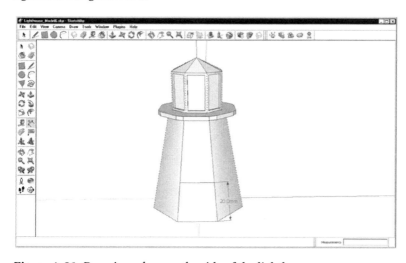

Figure 4–21. Drawing a box on the side of the lighthouse

Create two lines starting at the top corner of the box and pointing out until it is at the outer edge of the base. Since the box in Figure 4–21 is on the green dashed axis, you want to make sure the line points in the same direction. As you draw the lines that point out, a diagonal dashed line will appear (Figure 4–22a). This line lets you know you have hit the edge of the base. The direction of your line will depend on the orientation of your model. Repeat the same process at the other corner of the box. Then create a line directly down to the corner of the base. Select the inner surface defined by the box, and hit Delete to erase the surface (Figure 4–22b). On both sides of the box you should now have overhanging triangles. Using the Line tool connect both triangles, as shown in Figure 4–22c.

a.

b.

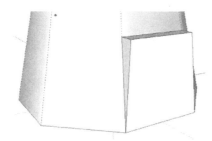

c.

Figure 4–22. Completing the box on the side of the lighthouse

The next step is to create a door in the model. To assist you in creating the door, you will be using guidelines. Select the Tape Measure tool. Select the bottom surface of the door, and drag your cursor upward. As you drag the cursor upward, notice the dashed line. Type **15mm**, and hit Enter on your keyboard (Figure 4–23).

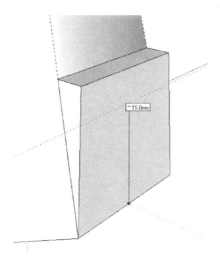

Figure 4–23. *Creating the door with the Tape Measure tool*

Create two additional guidelines that are 5mm toward the center from the sides (Figure 4–24a). Then using the Line tool, trace a path along the guidelines (Figure 4–24b). Using the Push/Pull tool, create an extrude 1mm inward (Figure 4–24c). You will be using the same approach to create all the windows around the lighthouse in the next section.

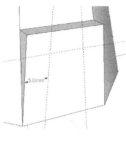

a.

b.

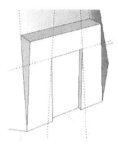

c.

Figure 4–24. Using guidelines to create the door

Creating the Tapered Wall Windows

The next step is to create tapered wall windows for the model. The model consists of seven windows. In this example, we will be creating one window, and then I'll leave it up to you to construct the rest. The process for creating all the windows is the same. The first window will be right above the door. Before you continue, delete all the other guidelines. With the Select tool, select the guideline, and press Delete on your keyboard. Select the Tape Measure tool, and create a guideline halfway between the top of the door entrance and balcony (Figure 4–25a). To make sure the guidelines stay on the surface of the model, follow a path along the edge of one of the surfaces you are placing the guideline on. Now create a second guideline 10 mm above the first guideline (Figure 4–25b). Create two additional guidelines from the side 3mm inward.

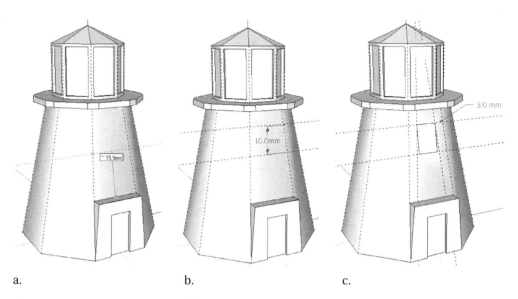

a.   b.   c.

Figure 4–25. *Constructing a window*

Using the Line tool, create a box within the guidelines (Figure 4–25c). The same way you created the surface in which you constructed the door, you do the same with the window, as shown in Figure 4–26.

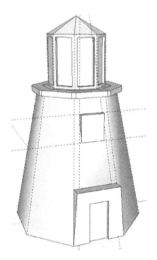

Figure 4–26. *Creating the surface for the window*

To add depth to the window as you did for the door, you would have to use the Offset tool and create a 1mm offset. Using the Push/Pull tool, create a 2mm inward extrusion. After adding all the windows to the model, you should end up with a design similar to Figure 4–27.

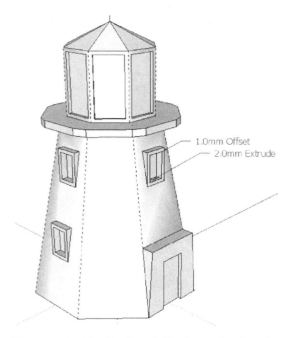

Figure 4–27. The final model before upload to Shapeways

Double-Check Before Uploading

Now that the lighthouse model is complete, it is almost ready to upload for 3D printing. But first you need to review the five rules mentioned earlier in this chapter:

- Is the model closed?
- Are the white surfaces facing outward?
- Is the model manifold?
- Does the model meet the specifications for the material?
- Is the model structurally stable?

You can review these rules in any order. But make sure to cover all of them before you upload your model for 3D printing on Shapeways. You may or may not encounter the same problems that I do. So, keep an eye open, and look at places within the model other than the ones shown as illustrations in this book.

Are all the White Faces Pointing Outward?

Using the Orbit tool, rotate around the model, and see whether you can find anything peculiar. With a close inspection of the lighthouse model, you'll notice one thing right away. Some of the surfaces of the model are facing inward. When designing a model in SketchUp, you will come across two surface colors:

a white surface and a gray surface. If you manually changed the color of the model, then the surface colors might be different. The white surface of the model needs to be facing out from the model.

Do you see any gray surfaces facing out of the model? In Figure 4–28a, you can see that some of the surfaces that make up the window are gray. Right-click each gray surface in the model. A drop-down menu will appear; select Reverse Faces (Figure 4–28b). The surface will change direction, and now you should see the white surface facing outward. If your model has a lot of gray surfaces facing outward, then to speed the process of flipping the surfaces, you can use Orient Faces. Select any of the white surfaces in the model, and right-click. From the drop-down menu, select Orient Faces. Now all the surfaces in the model will orient themselves in the direction of the selected face.

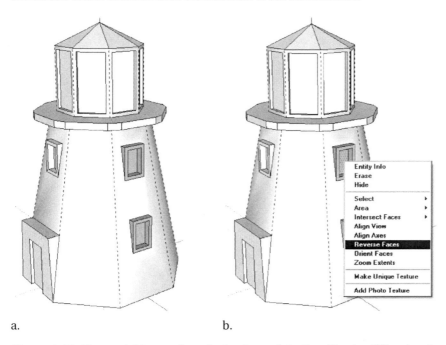

a.   b.

Figure 4–28. The model has surfaces facing inward, indicted by the different color shades.

Is the Model Manifold?

The model from the outside might look normal, but there might be problems within the model that you cannot see. That's why you need the Section Plane tool. The Section Plane tool allows you to look into the model. From the Tools menu, select Section Plane.

A green square will appear on your cursor. Place the green square anywhere on the model, and it will orient itself to that surface. To place the section plane on top of the lighthouse, make sure to place the cursor on the topmost point of the roof (Figure 4–29a). Click to lock the section plane in place. The section plane will change color to orange, with shorter arrows (Figure 4–29b). Select the Move/Copy tool, and click the orange square. Move the section plane down, and notice the model is being cut away. As you go through the object, check to see whether there are any extra lines, changes in color, or extra surfaces. Extra surfaces in the model could result in a manifold error. From the top, as you cut away at the model, you'll notice there are few extra lines around the inner edge of the model (Figure 4–30), which was probably left behind during the construction of the model. These need to be deleted, or they will show up as errors in the model when uploading to Shapeways.

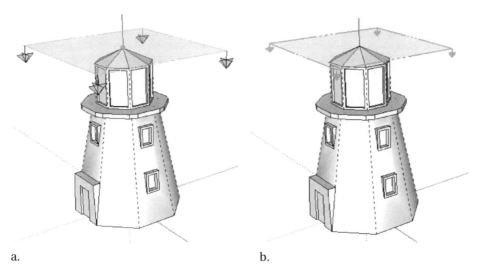

a.   b.

Figure 4–29. Slicing through the model using Section Plane

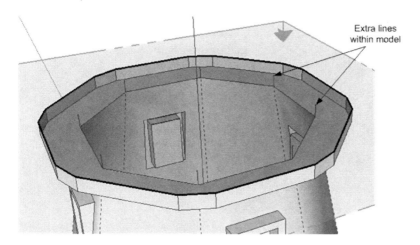

Figure 4–30. Extra lines found internal to the model

Using the Select tool, select each line, and delete it. Also, take a section plane of multiple sides. Sometimes it's difficult to catch a mistake from one angle. So far, you have not seen any hanging surfaces or internal surfaces that might be hiding inside the model. This is good news. Manifold errors are common, but it looks like we won't be having that problem. Once you are done with the Section Plane tool, select the section plane, and hit Delete on your keyboard. This will erase the Section Plane tool so you can now view your entire model.

Is the Model Closed?

Now you need to make sure there aren't any openings within the model. Using the Orbit tool, maneuver around the model, and search for any openings. A missing surface in the model indicates an opening. After close inspection, you'll find that the balcony has an open surface (Figure 4–31a). Using the Line tool, draw a diagonal line (Figure 4–31b). SketchUp will then autofill the surface.

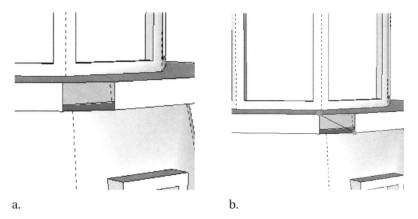

a.   b.

Figure 4–31. Filling in openings within your model

As your model get more complex, you are likely to encounter openings within the model. Just remember these openings need to be closed before they're uploaded. It's common to forget to check the bottom of the model. Make sure to look under the balcony for any problems also.

Does the Model Meet Specification, and Is It Structurally Stable?

The material of choice for 3D printing will be White, Strong & Flexible. The material is made of nylon and has good heat resistance. The minimum wall thickness for this material is .7mm, but Shapeways recommends having at least 2mm. Wall thicknesses smaller than 2mm could result in it being fragile.

You now need to double-check to see whether any part of the model has a wall thickness smaller than 2mm. After close inspection of the model, you'll notice that the windows have a depth of 1.9mm (Figure 4–32). Though this does not meet the minimum wall thickness of 2mm, you don't have to be too concerned about it since it is only a small portion of the entire model.

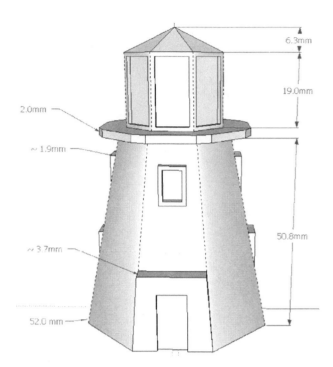

Figure 4–32. Dimensions of the lighthouse

The lighthouse has a bottom diameter of 52mm and a height of 78.2mm. Since most of the weight of the model is focused in the bottom of the model, structurally the model is stable.

Before you continue, remember to save the model. You can do a few additional things to improve the model. There is excess material within the model that you do not need. By getting rid of the material, you can reduce printing costs. For now, you will stay with the solid design. In Chapters 7 and 8, you will learn how to make the model hollow and reduce the costs of your design even further.

Upload for 3D Printing

Now let's upload the model for 3D printing. From the File menu, select Save, or press Ctrl+S on your keyboard. The Save As dialog box will appear. Save the file in a location for easy retrieval. Now to prepare your model for 3D printing, from the File menu select Export ➤ 3D Model. The Export Model dialog box will appear (Figure 4–33).

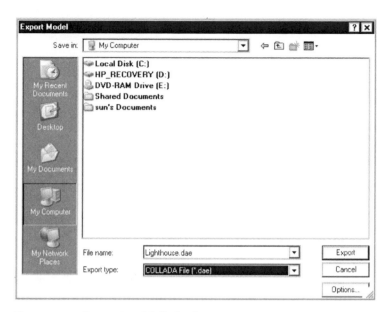

Figure 4–33. Export Model dialog box

Select COLLADA File as the export type, and pick a location to export the file. Name the export **Lighthouse**, and then click Export.

Now it's time to upload the Lighthouse.dae file to Shapeways for 3D printing. Open your Internet browser (Firefox or Internet Explorer), and browse to www.shapeways.com. On the Shapeways home page, select "upload" from the top menu bar. You will be directed to the Upload a New Model page (Figure 4–34). This is where you will be uploading the Lighthouse.dae file.

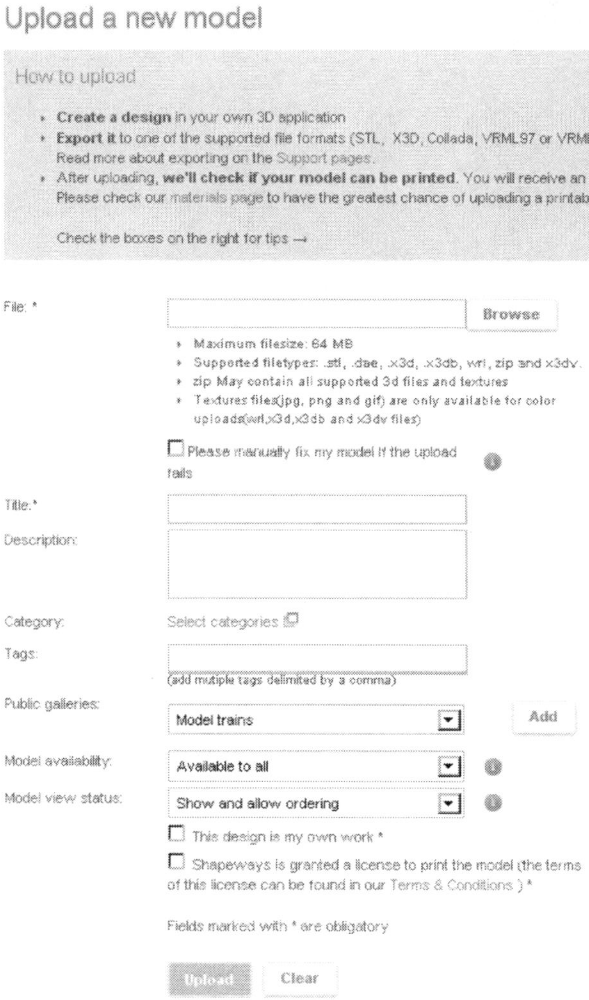

Upload a new model

How to upload

> • **Create a design** in your own 3D application
> • **Export it** to one of the supported file formats (STL, X3D, Collada, VRML97 or VRML2).
>   Read more about exporting on the Support pages.
> • After uploading, **we'll check if your model can be printed**. You will receive an email with the result.
>   Please check our materials page to have the greatest chance of uploading a printable model.
>
> Check the boxes on the right for tips →

File: *   [   ] [Browse]

> • Maximum filesize: 64 MB
> • Supported filetypes: .stl, .dae, .x3d, .x3db, .wrl, zip and x3dv.
> • zip May contain all supported 3d files and textures
> • Textures files(jpg, png and gif) are only available for color
>   uploads(wrl,x3d,x3db and x3dv files)

☐ Please manually fix my model if the upload   ⓘ
fails

Title: *   [   ]

Description: [   ]

Category: Select categories ⊡

Tags: [   ]
               (add mutiple tags delimited by a comma)

Public galleries: [Model trains   ▼]   [Add]

Model availability: [Available to all   ▼] ⓘ

Model view status: [Show and allow ordering ▼] ⓘ

☐ This design is my own work *

☐ Shapeways is granted a license to print the model (the terms
of this license can be found in our Terms & Conditions) *

Fields marked with * are obligatory

[Upload]   [Clear]

Figure 4–34. Shapeways model upload page

There are eight categories to fill in the form, as described in Table 4–1. The parts you have to fill out are denoted by an asterisk (*). The rest of the information is voluntary.

Table 4–1. Model Upload Form Options

Title	Description
File	The file types that Shapeways accepts for upload are `.stl`, `.dae`, `.x3d`, `.x3db`, `.wrl`, `.zip`, and `.x3dv`. The maximum file size for upload is 64MB.
Title	Enter a name for the model you want to upload.
Description	Give the model a description
Category	Select a category in which you want the models to be displayed in, such as Art, Gadgets, Games, Home décor, Jewelry, Hobby, and Seasonal.
Tags	Enter words that describe your model.
Public galleries	Select a gallery in which you want your model to be displayed.
Model availability	"Available to all" means it will be placed in the gallery for anyone to see and rate. "Only for myself" means only you can view it.
Model view status	With "Show only," you can only order the model. "Show and allow ordering" means anyone can order it.

For the lighthouse model, follow these steps:

1. Click the Browse button, and upload the `Lighthouse.dae` file.

2. For the title, enter **Lighthouse**.

3. Under Description, enter a few sentences describing the model.

4. For Category, select categories in which the model would be suitable for display.

5. For tags, enter **lighthouse, house, light**.

6. To place the model in the Public gallery, select the gallery from the drop-down list, and then click the Add button.

7. For "Model availability," select "Only for myself."

8. For "Model view status," select "Show and allow ordering."

After you have filled in the entire form, make sure to select the two check boxes at the bottom of the upload page: "This design is my own work" and the "terms & conditions." Once the information has been input into the system, click Upload. If there are no errors in the upload process, the upload usually takes less than a minute to complete. After upload, the model will show up in your "my designs" gallery (Figure 4–35).

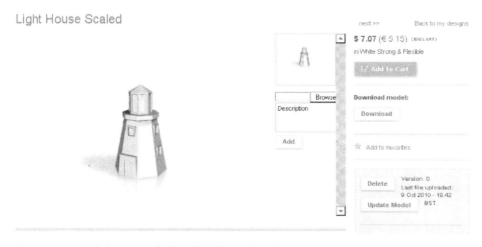

my designs

| Sort by | Upload date ▼ | Last to first ▼ | 5 rows ▼ | | 1 of 1 |

How to join the
Co-Creator Platform
1 2 3 in a few steps!

user details

Sandeep Singh

I'm a 3D modeler, writer, and I enjoy
building things

rav profile

Light House 2
$ 73.54 | (€ 53.58)
rapid
★★★★★ (0)

Bracelet V
$ 60.00 | (€ 43.71)
rapid
★★★★★ (0)

Ring Poem
$ 6.09 | (€ 4.44)
rapid
★★★★★ (0)

| Sort by | Upload date ▼ | Last to first ▼ | 5 rows ▼ | | 1 of 1 |

Figure 4–35. Shapeways "my designs" gallery

Wow! At first glance you might be surprised to see that the model costs $73.54. The model only measures 52 mm (L) × 52 mm (W) × 78.2 mm (H). So, how can a model so small cost so much? Currently 3D printing costs are expensive, but as demands increase, I expect there to be cost reductions.

But for now there are two ways of getting around the high prices. The first method is to hollow out the inside of the model mentioned earlier, and the second method is to scale the model smaller. In Chapter 6, you will learn how to scale a model, and in Chapter 7, you'll learn how to hollow out a model. If you think you have an idea of how to scale the model and hollow it out, give it a try. Hint! You will need to use the Scale tool.

After a few modifications, I was able to reduce the price of the model to $7.07. The entire model in Figure 4–36 was scaled so the base measured 25.4mm instead of 52mm. Scaling the model down shrunk the width of the balcony to 1mm. So, the width of the balcony was increased by 1mm. The model was also hollowed out from the bottom (Figure 4–37b).

Light House Scaled

next >>   Back to my designs

$ 7.07 (€ 5.15) (INCL VAT)
in White Strong & Flexible

[☐ Add To Cart]

[Browse]
Description

Download model:

[Download]

[Add]

☆ Add to favorites

Delete	Version: 0
	Last file uploaded:
	9 Oct 2010 - 19.42
Update Model	BST

Figure 4–36. Lighthouse scaled and hollowed in Shapeways

To order the model, make sure to select the model from the "my designs" page. On the model page, click Order. The model measures 4cm (H) × 2.3 cm (W) × 2.3 cm (L). Within two weeks, the model will arrive in the mail. Figure 4–37 shows the final results of the model.

 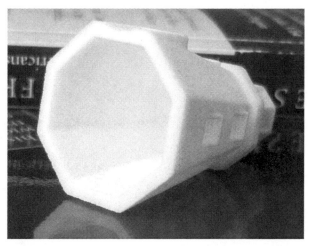

a.   b.

Figure 4–37. 3D printed lighthouse model

Summary

This chapter was all about taking what you learned in Chapters 1 through 3 and applying it to design a model in Google SketchUp for 3D printing in Shapeways. You learned how to create a mind map, construct premodel sketches, model in SketchUp, and upload the design for 3D printing to Shapeways. In the next chapter, you'll learn to install plug-ins and how they can be used to increase your productivity. You'll also look at the Outliner, an organizational feature in SketchUp.

■■■

CHAPTER 5

Tools and Techniques to Save Time

In this chapter, you'll explore a few tools and techniques you can use to save time and increase productivity as you model. As you might have noticed in Chapter 4, the models you design can very easily get complicated with multiple parts and sections. To start the chapter off, you'll explore the Outliner, a built-in feature of SketchUp for organizing models.

Halfway through the chapter, you'll switch gears to learn about SketchUp plug-ins. You'll learn how to use the Shapes, Volume Calculator 21, Flattery Papercraft, and CADspan plug-ins. You don't have to worry about purchasing any of the plug-ins, because they are free for download—isn't that great? Let's get rolling!

Organizing Your Models

One of the most important features of SketchUp is its ability to organize your models. With the organizational tools in SketchUp, you can isolate and view specific parts of your model without having to manage the entire model simultaneously. Organization improves the visualization, editing, and presentation of your models. Most importantly, if the model is well organized, you can save time and solve any problem effectively.

Without installing any plug-ins, right of the bat you can use some built-in features that come with SketchUp to help you organize your models. These are the Outliner, Make Group, Make Component, and Layers Manager tools. In the next section, we will go through each, demonstrating how they can be used during the modeling process and while organizing models. These tools are important when you design anything that has many parts, such as a car or furniture in a house. They are also very useful when your model has a lot of copies of the same part.

A Quick Look at the Outliner

The Outliner is where all the details of your model are stored. Think of it as a map showing where each part in your model goes. Right now, if you take a look at the Outliner dialog box, which is accessed through the Window ➤ Outliner, you will notice that it is empty, except for the term Untitled (Figure 5–1). This is because the modeling window in SketchUp is empty.

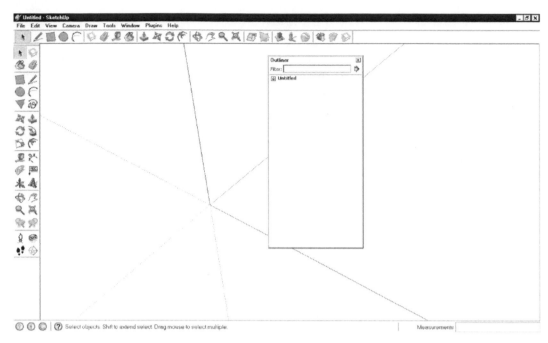

Figure 5–1. Outliner dialog box, empty

As you make a group, make a component, add a model, or add a layer in the Modeling window, the Outliner dialog box will start organizing everything into a hierarchical tree. By clicking each entry in the Outliner window, you can then access the different groups, components, and parts of the model.

What Is a Group?

Groups in SketchUp are used to combine all the parts of a model. All the surfaces and edges are grouped into a single entity that you can easily move or copy. If models were not grouped, you would have to first select all the surfaces before making a copy or moving them. To create a group, select all the parts of the model you want to combine, and then right-click and select Make Group.

What Is a Component?

You create a component when you will be using multiple copies of a part in a model. The great thing about a component is each copy you make is an instance of the original. So, any changes made to one of the instances will automatically be updated into the other copies. To create a component, select the parts of the model you want to combine, and then right-click and select Make Component.

What Is a Layer?

Create a layer in your model if there are certain parts of the model you want to hide. For example, you might want to hide a house you modeled in SketchUp and display only the furniture within the house. You can also use layers to assist you in hiding parts of a model that is interfering with your design. To activate the Layers Manager, select Window ➤ Layers from the menu bar.

Using the Outliner

Let's add a couple of shapes to the modeling window and observe the changes in the Outliner. In the Outliner, you will see the hierarchical tree structure that makes up each of the boxes when you first draw them.

To get started, follow these steps:

1. Draw three boxes as shown in Figure 5–2. I have labeled the boxes 1, 2, and 3 for easy reference (you don't have to label them). Follow the steps to see whether you can create a similar structure in the Outliner.

2. After you draw the models, you must then identify them as groups, components, or layers. Select Box 1, right-click, and choose Make Component.

3. From the Window menu, select Layers. The Layer dialog box will appear (Figure 5–2b). Click the + sign to add a new layer. Select Layer1, and then draw the second box. Make sure the Visible box is selected and that the radio button is clicked on Layer1.

4. Select the Layer0 radio button, and then create a third box. Select the box, right-click, and choose Make Group.

   Once you are all done, your Outliner should look like Figure 5–2a.

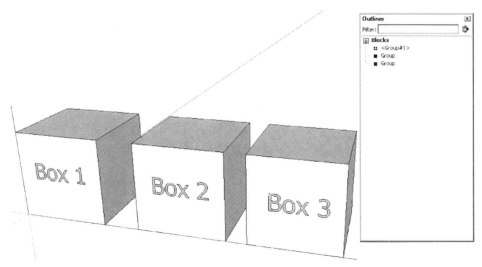

a.

b.

Figure 5–2. Three boxes and the Outliner

Now you can control each of the boxes within the Outliner. The first box is grouped as a component so you see four small squares and <Group#1>. The second and third boxes are groups and in the Outliner are denoted by a solid square.

5. Right-click the first group in the Outliner. From the drop-down menu, select Hide. The first box will be hidden from display. Try to hide the other boxes to see whether you can produce a similar effect.

6. Open the Layers dialog box again, and deselect the Visible check box for Layer1 (Figure 5–3a). Notice that Box 2 disappears. It's not actually gone. It's still on Layer1, but Layer1 is now hidden. To hide Boxes 1 and 3, select Layer1, and deselect the Layer0 radio button. Now Box 2 should be visible (Figure 5–3b).

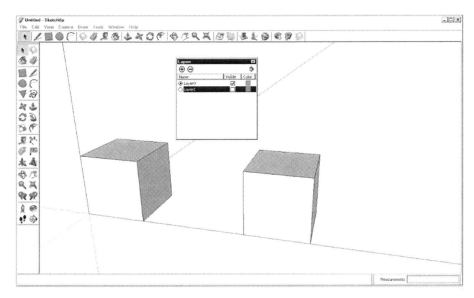

a.

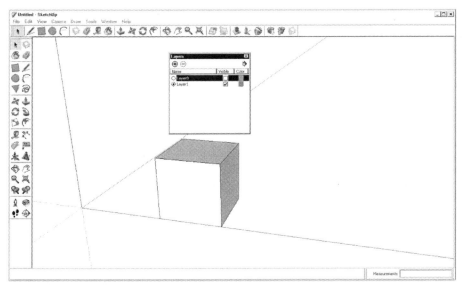

b.

Figure 5–3. (a) Layer1, deselected; (b) Layer0, deselected

What else can you do within the Outliner? Right-clicking each entity within the Outliner will present you with a set of options (Figure 5–4).

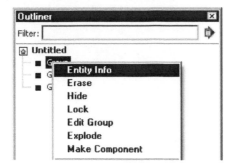

Figure 5–4. *Outliner options*

- Entity Info allows you to choose additional options such as Layers, Hidden, Locked, Cast Shadows, and Receive Shadows.

- Erase deletes the group or component in the model.

- Hide hides the group or component in the model.

- Lock disables the ability to move the object.

- Edit Group allows you to edit the object.

- Explode breaks your model into its individual entities.

- Make Component converts each group into a component.

We have just gone through a simple example showing how you can use the Outliner to organize and control the visibility of your model. In the next section, you'll look at a model that has several parts and see how the Outliner is structured.

Using the Outliner with a Complex Model

In the previous section, you saw an example of how the Outliner looks when you are working with a very simple model. But things can get quite complicated if the model has a lot of parts to it. In this example, you will look at a table that has several parts so you can better understand the power of the Outliner. Figure 5–5 shows a model of a desk; this is based on the desk that I'm using in my study. All of the parts that make up the desk have been divided into groups and components. Rather than have you reconstruct the model and the groups and components, you can simply download a copy of the model from the Apress Catalog page for this book. Look in the Chapter 5 folder for a file titled Grouped table. Open the grouped table in SketchUp. Once it's open, you should see a model similar to the one shown in Figure 5–5.

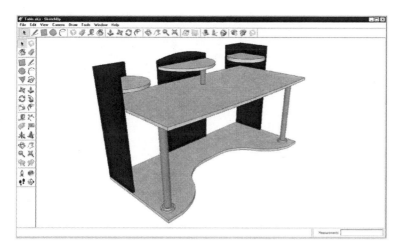

Figure 5–5. Desk model

When you look at the Outliner box for this model as shown in Figure 5–6, all the components in the model have been divided into sets: Back Support, Legs, and Horizontal Shelf. When you created a group of the box in the previous section, you selected all the surfaces and edges that made up the box and grouped it. With the table in Figure 5–6, you first grouped all of the individual sets into components and then made a group of the components. The solid square indicates a group. The checkered squares indicate components and subcomponents. You can create a component within a component to organize your models also.

1. Click the Back Support drop-down list. What you should see are the components that make up the back support of the table.

2. Select each component one by one within the back support, and notice those components are highlighted on the table.

To access all the components in the model, double-click the group in the Outliner. If you want to edit a component, then double-click it. Once within a component, you can edit it using any of the tools in SketchUp.

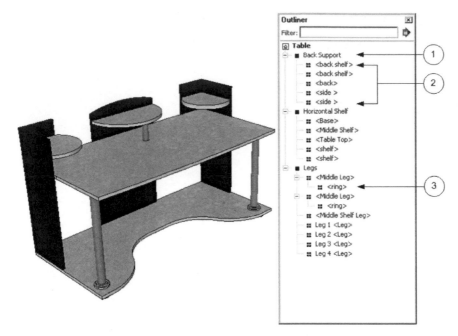

Figure 5–6. *To the right of the model is the populated Outliner dialog box.*

Hiding Groups and Components from Within the Outliner

If you were creating an instruction manual for putting together this table, then using the Outliner would be a great option. All you would need to do is create all the parts of the table and place them into groups or components.

1. To hide the back support of the table, right-click, and from the drop-down menu select Hide. If you just want to hide a single component within Back Support Tree list, right-click the component, and from the drop-down menu select Hide. Figure 5–7a shows the model's back support hidden, Figure 5–7b shows the horizontal shelf hidden, and Figure 5–7c shows the legs hidden.

    If you wanted to hide a select set of components in a group, hold down the Ctrl key, and select the components. Right-click and select Hide. Now only those components you have selected will be hidden.

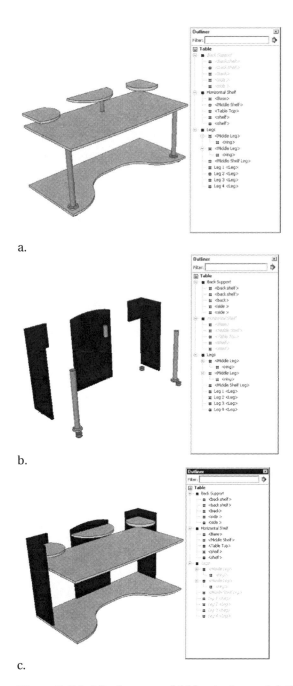

a.

b.

c.

Figure 5–7 (a.) Back support hidden in the model; (b.) all of the horizontal surfaces hidden in the model; (c.) all the legs hidden in the model

This process of organization within a model can greatly reduce modeling and debugging time. It also prepares the model for easy presentation to the customer and upload for 3D printing to Shapeways. You can easily delete groups and components and upload only those parts that you want to develop. In the next section, I'll go over a few plug-ins you can utilize to increase your overall productivity during the modeling process and before sending your model off for 3D printing.

Working with Plug-ins

A plug-in is a program that works within a software application, in this case Google SketchUp, to execute a function. Plug-ins allow the user to enhance modeling functionality, reducing modeling time and easing modeling frustrations.

Online you can find an assortment of plug-ins free for download. A search for the keywords *SketchUp plug-ins* will present you with an abundance of plug-ins to choose from. Table 5–1 provides a list of some of sites to visit.

Table 5–1. *List of SketchUp Plug-in Sites*

Website	Description
www.sketchuptips.blogspot.com	This is a site called Jims SketchUp [Plugins] Blog. Here you will find an assortment of plug-ins to download as well as resources for Google SketchUp.
www.sketchup.google.com/↵ download/plug-ins.html#lightup	Google SketchUp has put together a list of plug-ins that you can download.
www.alexschreyer.net	Alexander C. Schreyer, a PhD student at the University of Massachusetts, has put together a great set of plug-ins. Check out some of his other projects and video tutorials as well.
www.smustard.com/scripts/	This site offers a collection of plug-ins developed by the SketchUp community.

In the next couple of sections, you'll look at a few plug-ins that can assist you while you are modeling and a few plug-ins you can use to predevelop your models even before sending them for 3D printing: Shapes, Volume Calculator, Flattery Papercraft, and CADspan. So, let our adventure begin.

Shapes Plug-in

You will find the Shapes plug-in useful at cutting down your modeling time when you have to design geometric shapes. Developed by @Last Software, Inc., you can use the Shapes plug-in to model boxes, cylinders, cones, toruses, tubes, prisms, pyramids, and domes with only a few clicks of your mouse. Before you get started and see how the model works, you will need to download the plug-in and place it the Plugins folder of Google SketchUp. To access the Plugins folder, you need to follow a path similar to this C:\Program Files\Google\Google SketchUp 8\Plugins. It might be different depending on the computer you are using. You can download a copy of the plug-in from http://sketchup.google.com/intl/en/download/rubyscripts.html. The plug-in requires that you also install parametric.rb and mesh_additions.rb, which are available on the site.

1. Once you download and place the three files into the Plugins folder, open SketchUp.

2. From the menu bar in Google SketchUp, select Draw ➤ Shapes. You will then be presented with a list of shapes to choose from (Figure 5–8a).

3. When you select a shape from the list, a dialog box will appear asking for the parameters of the model you want to create (Figure 5–8b). Enter them, and click OK.

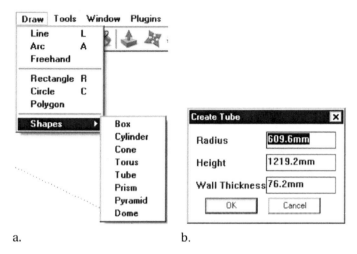

a.   b.

Figure 5–8. (a.) List of shapes; (b.) enter the dimensions on the shape.

Figure 5–9 shows models of the shapes created with the Shapes plug-in.

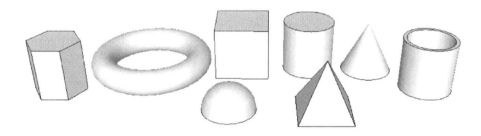

Figure 5–9. A pyramid, prism, dome, tube, box, cylinder, cone, and torus—all modeled with the shape.rb
plug-in

Wasn't that easy to model? If you were to draw these shapes from scratch using the tools in SketchUp, it would take you a lot longer. You have saved a lot of time, so now you can work on other parts of the model. Do you see the benefit of using a plug-in in SketchUp? Well, continue reading, and I'll show you couple more plug-ins I'm sure you will find useful.

Volume Calculator Plug-in

Knowing the volume of a model can be very important for determining its price. As you have noticed on Shapeways, the cost of the model depends on the amount of material you use. Using Volume Calculator 21, developed by TGI, you can easily calculate the volume and make adjustments to your models. This tool is especially helpful when you're working with complex volumes. Let's go through the basic steps of using Volume Calculator 21. You can download a copy of the plug-in from `www.cad-addict.com/2008/11/sketchup-plug-ins-volume-calculator.html`.

1. Right-click, select the `VolumeCalculator21.rb` link, and from the drop-down list select Save Link As. Save the file in your Google SketchUp 8 `Plugins` folder.

    Before we dive in and do volume calculations with the plug-in, take a look at Table 5–2, where I have done some calculations based on the formula of each shape using the Constants in Feet values. The solution of each formula is shown under Hand Calculation. Use the table a reference to test the calculations from Volume Calculator 21.

Table 5–2. Volume Formulas

Shape	Formula	Constants in Feet	Hand Calculation
Cone	$V = \dfrac{1}{3}\pi r^2 h$	$r = 2, h = 3$	25.12 cu. ft.
Cube	$V = a^3$	$a = 2$	8 cu. ft.
Cylinder	$V = \pi r^2 h$	$r = 2, h = 4$	50.24 cu. ft.
Rectangular prism	$V = l \cdot w \cdot h$	$l = 2, w = 3, h = 4$	24 cu. ft.
Pyramid	$V = \dfrac{1}{3}ah$	$a = 2, h = 3$	2 cu. ft.
Sphere	$V = \dfrac{4}{3}\pi r^3$	$r = 2$	33.49 cu. ft.

2. If you have SketchUp open from the previous example, close and reopen the program. When SketchUp is open and you place a plug-in into the `Plugins` folder, SketchUp does not detect it. Only when you open SketchUp does it check all the files within that folder.

3. Using the Line and Push/Pull tools, create a cube that has a length, width, and height of 2 feet. Remember to enter 2 feet; in SketchUp, you would use **2'**.

4. After you draw the cube, make it into a group, or else Volume Calculator 21 will not work. To do that, highlight the entire model, and right-click it. From the drop-down menu, select Make Group. When you select the model now, it will be surrounded by a blue box.

5. Right-click the model, and from the drop-down menu select Volume. The Volume Parameters dialog box will appear (Figure 5–10).

6. To calculate the volume of the cube, select Units as cu.ft, because you will be using cubic feet. The Layer option lets you choose the layer, Hide Edges presents you with the edges, and Color allows you to set the desired color for the job.

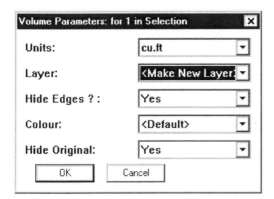

Figure 5–10. Volume Parameters dialog box

7. Once all the selections have been made in the Volume Parameters dialog box, click OK. The calculated volume of the cube should be 8 cu. ft. Compare your results with the hand calculations in Table 5–1. If the results do not match, try drawing the model again. This time, make sure the constants shown in Table 5–1 are used when modeling the cube. Figure 5–11 shows the cube modeled in SketchUp. Try to see whether you can get similar results with the other shapes in Table 5–2.

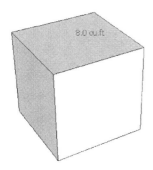

Figure 5–11. The volume of a cube (8 cu. ft.)

Flattery Papercraft Tool

Flattery is a 3D imaging plug-in developed by Google. Flattery is a unique plug-in in that you can use it to unfold any of your models on 2D planes. This plug-in is especially useful to 3D designers when they need quick mock-ups of their designs for demonstrations and feedback. Having a design that you can easily print with your personal desktop printer is much easier and cheaper than sending it out for 3D printing. On top of all that, it is free for download. You can download a copy of Flattery from www.pumpkinpirate.info/flattery/. Unzip the download, and place the files in the Flattery folder with your Google SketchUp Plugins folder on your computer. Once you have placed the files, remember to reopen SketchUp for the install to take effect.

Once Flattery is installed, SketchUp will display the Flattery toolbar in SketchUp. The toolbar consists of five buttons: Index Edges, Reunite Edges, Add Tabs, and SVG Export (Figure 5–12).

Figure 5–12. Flattery toolbar

Using the Flattery plug-in is very simple. First you will need a model to unfold. So, construct a pyramid in SketchUp. You can choose any dimension you like.

1. Now select the entire pyramid, and click the Index Edges button.

    This will create an index of all the edges in your model so that Flattery knows which edges were connected for the Reunite Edges tool.

2. Now deselect the entire pyramid in SketchUp, and then click the Unfold Faces button.

    To deselect the model, simply click in an open space in the modeling window before clicking the Unfold Faces button.

3. Click any surface in the model, and then click a neighboring surface.

    Each surface in the model will unfold and align itself with every new surface you click (Figure 5–13a). Continue selecting all the surfaces in the model until all of them have been unfolded (Figure 5–13b). The next two steps are optional; you can skip them and continue with step 7.

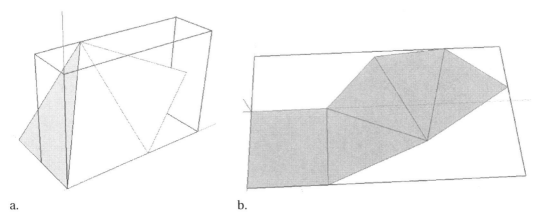

a.   b.

Figure 5–13. Unfolding of the surfaces using the Flattery plug-in

4. With the Selector tool, double-click to enter the grouped surfaces, and select the Reunite Edges tool.

   You can use the Reunite Edges tool to readjust the location of each surface in the model.

5. Click the surface or surfaces in the model that you want to move. Place the cursor over one of the edges of the model, and the corresponding edge on the selected surface will appear in red (Figure 5–14a).

6. Click the edge, and the surface will attach to the selected edge (Figure 5–14b).

This is a great tool to use especially when you have a complex model with many surfaces.

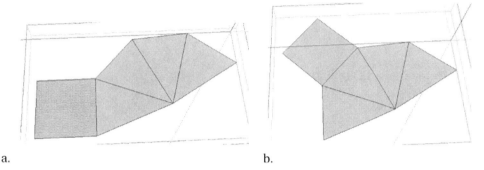

a.   b.

Figure 5–14. Using the Reunite Edges tool to move surfaces

Next you'll add tabs to the surfaces.

7. Select the Tab tool, and place the cursor over an edge.

   Its corresponding edge will appear red.

8. Click the edge once, and drag the mouse to view an outline of the tab being created (Figure 5–15a). Click again once you are happy with the outline to create the tab.

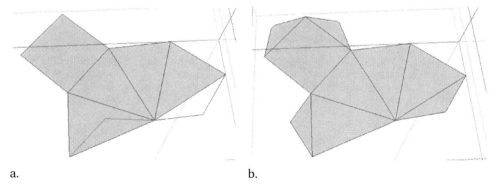

a.   b.

Figure 5–15. Creating the tabs for your model

9. Continue the process to draw in all the tabs in the model.

   All that is left for you to do is export the model for printing.

10. Select the SVG Export tool, and save an export of the file onto your computer.

You are all done; now, using a photo-editing program such as Inkscape or Photoshop, you can easily print the design onto paper to be cut out and fold. As mentioned earlier, you can unfold your 3D models and print them effortlessly, saving time and money. This is a great way to test your design before sending it off for 3D printing.

CADspan Plug-in

CADspan is great for converting your models from SketchUp files with the .skp extension to an STL file, especially if you need to upload your file for 3D printing to another service other than Shapeways. In Chapter 12, I discuss a few companies that use STL files for 3D printing their models. On top of being a file converter, CADspan can be used to debug your models for any problems. In addition to converting models to STL files, the plug-in has features such as Resurface, Layerize, Unsmooth Model, Preview Style, Import Geometry, and Export to Raw STL (Figure 5–16).

Figure 5–16. *CADspan dialog box in SketchUp*

So, what does each tool do in SketchUp?

- Resurface creates a mesh around the model to convert the model into a solid structure.

- The Layerize tool allows you to select the entire model or surfaces and place them on a layer.

- The Unsmooth tool is used to remove smooth surfaces from the model.

- Preview Style highlights all the problematic areas in the model with a red color.

- Import Geometry allows you to import STL files for checking before uploading to CADspan.

Before you see how you can use CADspan in assisting in debugging and converting your models, you will first need to download a copy of the CADspan plug-in. To download your own copy, visit www.cadspan.com and on right of the web site click CADspan Plugin (Figure 5–17). Then on the next page, click to download the plug-in. This will direct you to the download page where you can select the version of CADspan to download. On a Windows computer, click Windows: Google SketchUp 8 to download the plug-in onto your computer.

Unlike other plug-ins you have installed, this plug-in downloads as an executable file. Double-click the executable file, and follow the on-screen instructions to install the plug-in.

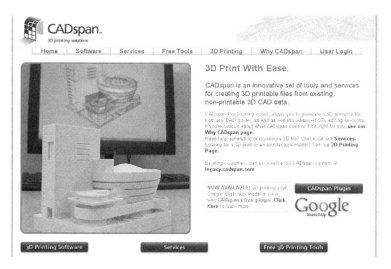

Figure 5–17. CADspan web site

After installation, open Google SketchUp. The plug-in will appear on your Getting Started toolbar.

Understanding the STL Format

Now that the plug-in is installed, you'll take the lighthouse model you designed in Chapter 4 and convert it into an STL file. But before doing that, let's briefly review what an STL file is.

The .stl file is a common 3D printing file format that you will encounter as you continue 3D modeling beyond just this book. STL files store triangulated information of 3D surfaces. Each surface is broken down into smaller triangles that are described by three points and a perpendicular direction. You can think of describing a 3D surface with a mesh of triangles. It is this mesh that describes the shape and form of the model. Think of the Eiffel Tower in Paris, France. It is built with a mesh of triangles.

An STL file is described by facets. Each facet consists of a triangle and a norm that is perpendicular to the triangular surface (Figure 5–18). Each of the vertices of the triangle is described by three data points v_x, v_y, v_z, and the norm is described by n_i, n_j, n_k. In total, there are 12 points that describe a single facet. Following the right-hand rule, with your thumb facing norm and fingers pointing in the counterclockwise direction, the vertices are all positive values.

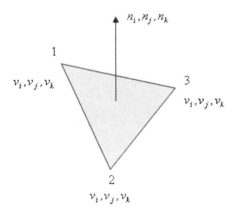

Figure 5–18. Twelve points that describe a single facet

You will encounter two types of STL file formats: ASCII and binary. Although ASCII STL files are more popular when testing or debugging a system, they are very big and impractical to use. Binary STL files are much more commonly used and far more practical than ASCII.

Converting to an STL File

Now you'll take the lighthouse model that you designed in Chapter 4 and convert it into an STL file.

    1.   Open the Lighthouse.skp file from Chapter 4, and select Preview Style.

           This will highlight the model in either of two colors, brown or red. If any of the outer surfaces of the model are red, then the surface must be reversed. In that case, reverse the surface so that you see brown on the outside. If you see a red surface facing outside, then one or more of your surfaces is facing outward. During 3D printing, this will appear as an error, and Shapeways will not upload your model. Right-click, and select Reverse Faces from the drop-down list to reverse all the red faces. You should now have a model like Figure 5–19.

Figure 5–19. Lighthouse model to preview style

2. Now select Resurface, and the CADspan Resurfacer dialog box will appear (Figure 5–20a).

    If this is your first time opening the CADspan Resurfacer dialog box, then you will need to register for a user name and password. Click Register to create your user name and password on CADspan.

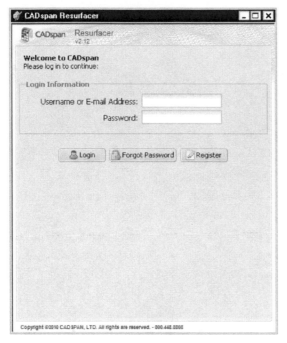
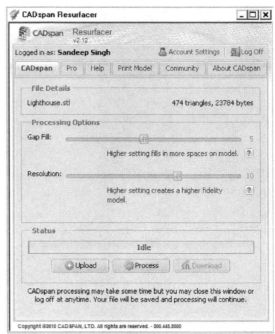

a.   b.

Figure 5–20. CADspan Resurfacer

Once logged in, the CADspan Resurfacer menu options appear (Figure 5–20b).

1.   Within Resurfacer are six tabs; select the CADspan tab. Under CADspan are File Details, Processing Options, and Status. There is a 7,500 polygon upload limit for the free version of CADspan. In the center of the CADspan window, you will see two sliders: Gap Fill and Resolution. Gap Fill is used to fill in holes in the model. Higher settings are recommended for models with many holes. Lower settings are recommended for models with fewer holes. Too high or too low of a setting could result in webbing or could create leaks in the model. Higher-resolution models will produce better-looking .stl files. For this example, you will stay with the default settings.

2.   Click Upload to upload the model to the CADspan server, and then select Process. CADspan then converts the design to an .stl file format. The process takes a few seconds to a few hours depending on the complexity of the model. Once processing is complete, the progress bar will read Complete. The lighthouse model uploaded has 474 polygons/triangles. Then click Download.

3.   The File Download box will appear (Figure 5–21). Click Save to save the zip file on your computer. Unzip the file after download to access its contents.

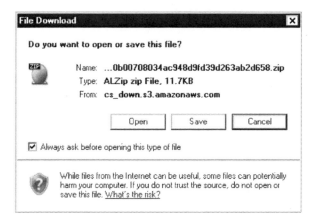

Figure 5–21. File Download dialog box

Within the ZIP file will be the STL file. You are now ready to upload the file to Shapeways or any other 3D printing service that accepts STL files. Check out Chapter 12 where I go over a couple of other 3D printing services that will accept STL file uploads.

Summary

In this chapter, you learned about the Outliner and how you can use it to organize models within SketchUp. You also learned about groups, components, and layers and saw how each can be used to organize a model and speed up modeling time. Then you explored a few plug-ins and saw how they can be a great resource when working with models in SketchUp. In the next chapter, you will learn all about modeling curved shapes and applying groups and components to model a sundial.

CHAPTER 6

■ ■ ■

Breaking the Barrier

In this chapter, you'll design two models. The first model you will design is a chess pawn where you will learn the techniques of modeling objects with curved surfaces for 3D printing. In the second half of the chapter, you'll switch gears and learn how to design a sundial. By designing the sundial, you will learn how to use groups, components, and shadow settings when developing and testing the model.

Designing Curved Models

You must be wondering by now how on Earth you could model that jar, cup, bottle, lamp shade, or ceiling fixture in your house using SketchUp. So far, you have modeled objects that have flat surfaces. Well, there is a way to design objects that have curved surfaces in SketchUp. The tools you will need to use are Follow Me and Arc. In this section, I'll demonstrate the steps of designing a chess pawn using both tools. If you were designing a lamp shade or bottle, similar methods would apply.

Creating the Pawn Template

To construct the pawn, you first have to change the view of the modeling window. From the menu bar, select Camera ➤ Standard Views ➤ Front (Figure 6–1). Once in the front view, you are ready to start modeling.

The pawn piece that you will be designing is based on the original Staunton chess piece. From rough measurements, the pawn has a width of 14mm and a height of 28mm. Since you already know what you are modeling, I have skipped the steps of developing a mind map and sketches of the model. I recommend you develop a few sketches of your model if you decide to model something different.

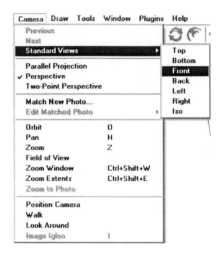

Figure 6–1. *Changing the view of your modeling window*

Starting from the center of the axis, follow these steps:

1. Select the Rectangle tool, click the axis, and then type **7mm, 28mm**.

   You will create only half the model and then copy the rest to the other half (Figure 6–2a).

2. Using the Tape Measure tool, create guidelines from the bottom of the box at 2mm, 6mm, 7mm, 8mm, 19mm, 20mm, and 21mm in height (Figure 6–2b).

3. Once the guidelines are in place, draw the outer edge that defines the pawn (Figure 6–2c).

   You can use the template in Figure 6–2c to guide you in constructing the edges of the pawn piece or observe an actual Staunton chess pawn online.

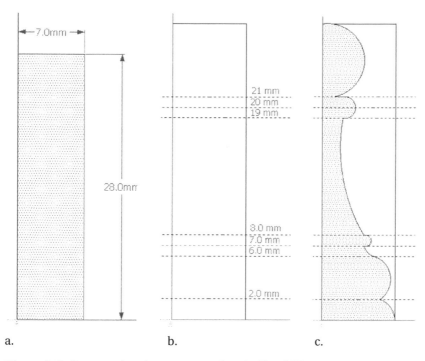

a.   b.   c.

Figure 6–2. Constructing the pawn template in SketchUp

4. Select the Arc tool, and then click the bottom-right corner and then the first guideline from the bottom at 2mm. Drag the cursor outward to create an arc. Click once more to lock the arc in place.

Continue the same process to create all the other arcs that define the pawn, as shown in Figure 6–2c. Once you have drawn the entire outer edge of the model, make sure there are no broken lines.

5. Select the surface on the left side of the arc.

If the left half is highlighted with blue dots, then there are no broken edges in the model. If not, then the line must be broken some place along the curved path you created. In this case, you need to zoom into the curve. Select the Zoom tool, and press and hold the left mouse key in the modeling window. Move the cursor left to right to zoom in and out of the model.

6. Once the broken lines are fixed, select and delete all the edges and surfaces to the right of the model.

You do not need the guidelines either. You can delete them or hide them in case you need them later. Click each guideline, and from the drop-down menu, select Hide. All you should be left with now is an outline of half a pawn (Figure 6–3, without the circle at the bottom).

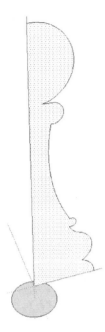

Figure 6–3. *Creating a circle below the model*

7. Below the model, draw a circle not part of the model but with its center along the blue axis.

    You will need to use the Orbit tool to tilt the model before creating the circle.

8. After creating the circle, select the model and the edge of the circle with the Follow Me tool. Rotate the Follow Me tool around the circle from start to finish, and then click at the end once more. This will create an outer shell of your model.

    If the model looks incomplete, then you will need to enlarge your template. Creating a curved path at a small scale in Google SketchUp does not work too well (Figure 6–4a), but there is a trick that will solve this problem. Using the Scale tool, you can enlarge the model and then apply the Follow Me tool. This seems to do the trick. Before you continue, press Alt+backspace to undo the previous step.

9. Select the entire model and then the Scale tool. Select the green box in the upper corner of the model, and type **10** (Figure 6–4b). Then, using the Follow Me tool, repeat step 8.

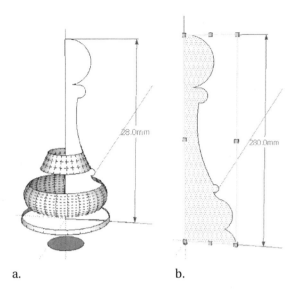

a.   b.

Figure 6–4. *Creating the pawn model with the Follow Me tool*

Your model should now look like Figure 6–5a. Now shrink the model to its original size.

10. Select the entire model with the Scale tool, type **.1**, and then press Enter.

The model was scaled 10 times its size in step 9. You are now scaling it back to its original size (Figure 6–5b).

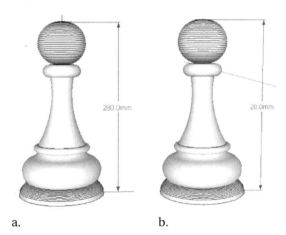

a.   b.

Figure 6–5. *Complete model of the pawn*

Double-Checking Your Work

You just learned how to create models with curved surfaces in SketchUp. Before you upload your model, remember that you need to double-check it for any errors. Do you still remember the five rules that were mentioned in Chapter 4? In case you have forgotten, I have listed them for your convenience here:

- Is the model closed?

- Are the white surfaces facing outward?

- Is the model manifold?

- Does the model meet the specifications for the material?

- Is the model structurally stable?

I'm assuming you are pretty familiar with the first, second, and last rules by now, so I will leave it up to you to double-check them. Now I want to focus my attention to the third and fourth rules. These rules are usually overlooked when designing models and are not visually/easily noticeable either.

Is the Model Manifold?

Select the Section Plane tool in the Tools menu, and create a section plane through the middle of the model (Figure 6–6a). Make sure there are no extra surfaces within the model. Though no extra surfaces are visible in the model to illustrate what one might look like, see Figure 6–6c. If you do see an extra surface in the model, select and delete it.

Now right-click the section plane, and from the drop-down menu, select Reverse (Figure 6–8b). You now will see the other half of the model. This is much easier than drawing another section plane. Now take a look at the model from the other side to see whether you find any nonmanifold errors.

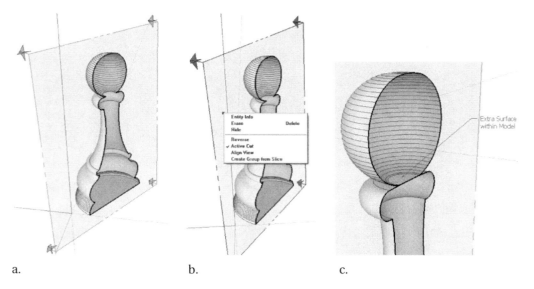

a. b. c.

Figure 6–6. Section plane cut

Adding Multiple Section Planes

Models that are more complex will require further observation. So, you might want to use multiple section planes at once while getting a better perspective of the model. To do that, you will have to group the first section plane to the model before applying the second.

Select the entire model, and right-click it. A drop-down menu will appear. From the drop-down menu, select Make Group (Figure 6–7a). Select the Section Plane tool, and create your first section plane on top of the model (Figure 6–7b).

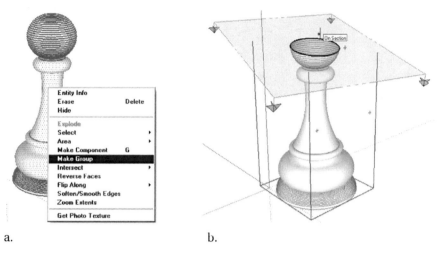

a.   b.

Figure 6–7. *Creating multiple section planes in SketchUp*

Now select the entire model along with the section plane. Right-click both, and from the drop-down menu, select Make Group. Using the Section Plane tool, create a section plane on the model but this time from the side (Figure 6–8).

Figure 6–8. *Section plane from the top and side of the model*

117

With two section plane cuts, you can easily view the model from all angles. This way, you can detect problem areas in the model and also make edits without having to delete one section plane and add another.

Does the Model Meet the Specifications for the Material?

Remember, Shapeways recommends that every model you design have at least a minimum thickness of 2mm. Anything smaller could make the model fragile. There are two areas in our pawn model that you need to check to make sure you meet this requirement. Using the Tape Measure tool, measure the diameter of the ball at the neck of the pawn (Figure 6–9a). Also make sure to measure the thickness of the stem (Figure 6–9b)

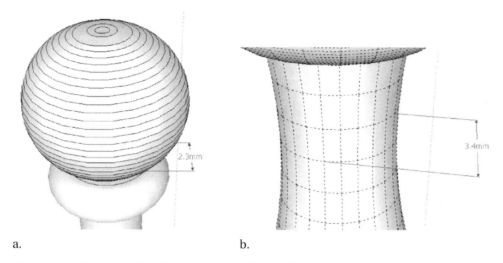

a. b.

Figure 6–9. Check whether the model meets the specification.

For this model, we meet the specification. The diameter of the circle between the ball and neck is 2.3mm, and at the stem it is 3.4mm.

Uploading for 3D Printing

The next step in the process is to upload the model to Shapeways. Export the model in the Collada file format for upload to Shapeways. Figure 6–10a shows the model after upload to Shapeways, and Figure 6–10b shows the model after being printed in 3D.

shapeways
passionate about creating
Pawn
Description
Title:
Main thumbnail image:
Default

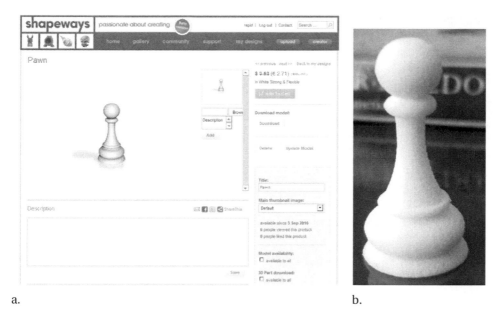

a. b.

Figure 6–10. *Pawn uploaded to Shapeways*

If the model costs more than $5, then you might want to check the dimensions of your model within Shapeways (Figure 6–11). If your model is larger than life, you might have forgotten to scale your model down in the "Creating Your Pawn" section.

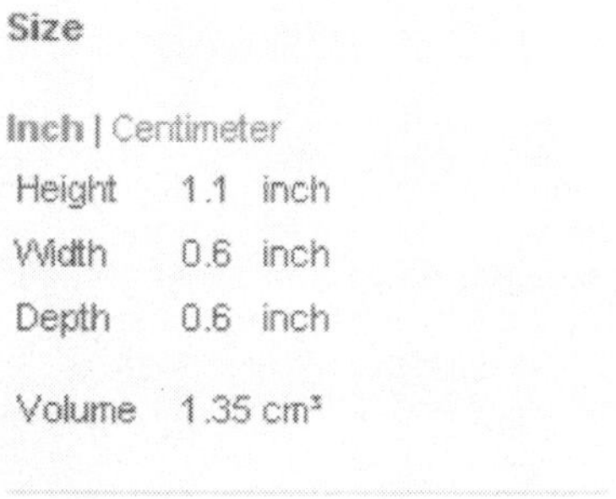

Size
Inch | Centimeter
Height 1.1 inch
Width 0.6 inch
Depth 0.6 inch
Volume 1.35 cm³

Figure 6–11. *The size of your model in Shapeways*

Designing a Sundial

In this section, you will be harnessing the powers of groups, components, and shadows to construct a model of a sundial in SketchUp. Unlike some of the other models you have constructed in this book, you will be applying some math to construct the sundial. Don't worry—the math isn't difficult at all. All the calculations will be conducted by an online application. The goal in this section is to have you design a functional model, in other words, a model that has real-world application.

Building the Theory

Sundials come in different forms and sizes and operate based on different mathematical principles. In this section, you will be constructing a horizontal sundial. Horizontal sundials can be designed with a horizontal line and a perpendicular line through the middle. The left side of the line represents 6 a.m., and the right is 6 p.m. The top of the perpendicular line is 12 noon, and the bottom is 12 midnight (Figure 6–12).

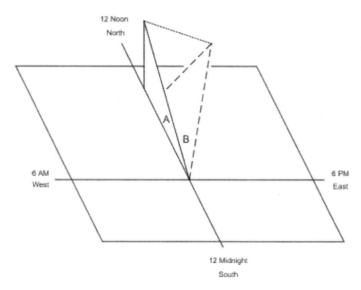

Figure 6–12. Principles of setting up a sundial

Since the sun is not visible during the night, all you care about is the time between 6 a.m. and 6 p.m. The solid triangle in Figure 6–12 represents the gnomon. The gnomon casts a shadow on the surface of the sundial denoted by the dashed triangle. The location of the shadow represents the time of day. In Figure 6–12, *A* represents the angle of the triangle, which is the latitude. *B* is the angle of the shadow created as a result of the gnomon. You can determine the angle of the shadow for each time period using the following formula:

$$\tan(B) = \sin(A)\tan(t \times 15°)$$

Solving for *B* in the formula will give you the angle for each time period in a 24-hour day. Using this information, you can then determine the angle of the hour hand for your sundial. But even before you can determine the angle of each hour, you will need to know the latitude at which the sundial will be placed.

What's Your Latitude?

For the location I'm in, it is 38 degrees. The latitude will vary depending on your location. To find your latitude, visit the Find Latitude and Longitude service at www.findlatitudeandlongitude.com. Select a location on the map to determine your latitude (Figure 6–13).

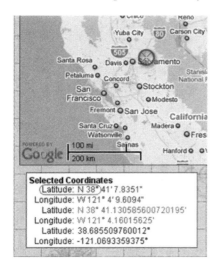

Figure 6–13. Find Latitude and Longitude service

Hand-calculating all the angles using the previous formula can be very time-consuming. Instead, you can use Professor Richard B. Goldstein's sundial calculator at www.providence.edu/mcs/rbg/java/sundial.htm (Figure 6–14).

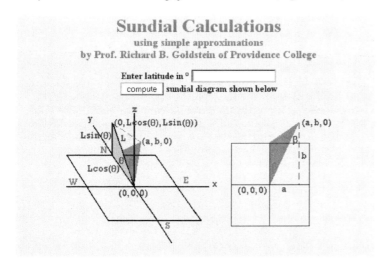

Figure 6–14. Richard B. Goldstein's sundial calculator

Enter the latitude that you found using the Find Latitude and Longitude web site (Figure 6–13) in degrees, and click Compute on Richard B. Goldstein's sundial calculator web site. The application will automatically calculate the angles for you. Scroll down the page to view the results. Notice that the calculations have been completed for the night hours as well (Figure 6–15). You will not be using the night results.

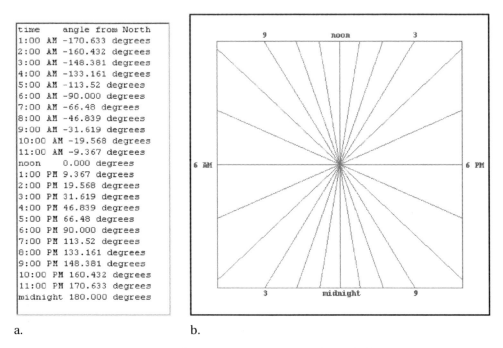

```
time       angle from North
1:00 AM  -170.633 degrees
2:00 AM  -160.432 degrees
3:00 AM  -148.381 degrees
4:00 AM  -133.161 degrees
5:00 AM  -113.52 degrees
6:00 AM  -90.000 degrees
7:00 AM  -66.48 degrees
8:00 AM  -46.839 degrees
9:00 AM  -31.619 degrees
10:00 AM -19.568 degrees
11:00 AM -9.367 degrees
noon      0.000 degrees
1:00 PM   9.367 degrees
2:00 PM   19.568 degrees
3:00 PM   31.619 degrees
4:00 PM   46.839 degrees
5:00 PM   66.48 degrees
6:00 PM   90.000 degrees
7:00 PM   113.52 degrees
8:00 PM   133.161 degrees
9:00 PM   148.381 degrees
10:00 PM  160.432 degrees
11:00 PM  170.633 degrees
midnight  180.000 degrees
```

a.   b.

Figure 6–15. Calculated angles from Richard B. Goldstein's sundial calculator

Drawing a Sketch of Your Model

As you have done in previous chapters, you will want to construct a few sketches of your sundial before modeling it in SketchUp. A sketch will keep you focused during the design process and also guide you through the design of the sundial. Figure 6–16 shows a model of the sundial you will be creating in this section. The sketch shows the front and back designs of the model. This will be a functional model that can be used as a watch, and I have gone one step further and added dimensions to the sketches. Because this sundial can be used in place of a watch on your hand, I have added 1mm pin holes where the watchband will go. The width of the sundial measures 6mm, and the length is 46mm.

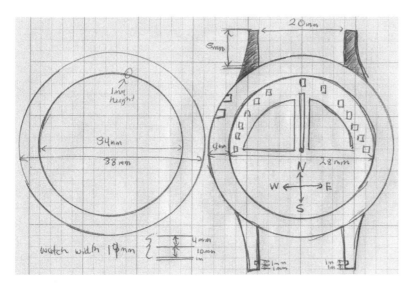

Figure 6–16. Sketch of sundial watch

Modeling in SketchUp

The first part of the model you will draw is the base, and then you'll work your way up in the design of the model. To start the modeling of the sundial, place the modeling window in Iso view.

Modeling the Base

To model the base, follow these steps:

1. Using the Circle tool, construct a 17mm radius circle starting from the center axis.

2. Extrude this circle down by 1mm (Figure 6–17a).

    It's easier to create this extrusion if you orbit to the bottom.

3. With the Offset tool, select the top surface of the circle, and drag it outward from the center so it is outside the circle. Then type **2 mm**, and finally hit Enter.

    A bigger circle will be created on top of the first (Figure 6–17b).

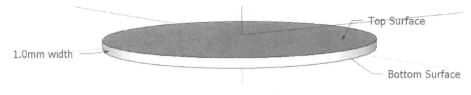

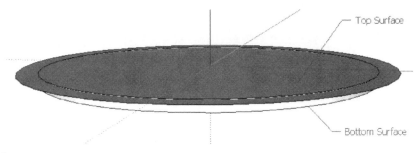

a.

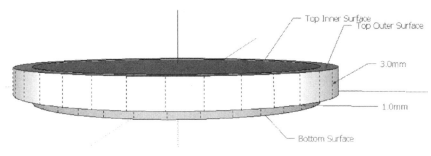

b.

c.

Figure 6–17. Creating the base of the model

4. Next extrude the top outer and inner surfaces by 3mm (Figure 6–17c).

    You will create an arc to blend in the two cylinders together. To do this, you use the Arc and Follow Me tools. Creating the arc will be easier once you activate the hidden lines that define each surface of the cylinder. Select View ➤ Hidden Geometry from the menu bar.

5. Zoom into one of the hidden lines, and using the Arc tool, create a line connecting the two bottom surfaces (Figure 6–18a).

    To complete the surface, draw diagonal lines from the sides to the inside of the circle (Figure 6–18b).

124

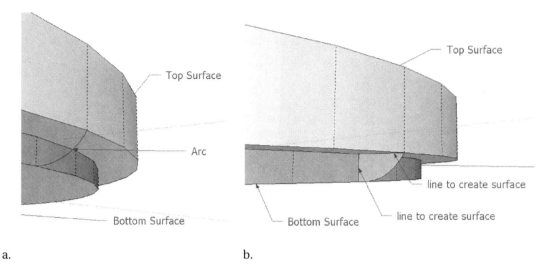

a.   b.

Figure 6–18. Creating the arc between the top and bottom surfaces

You will need to scale the model bigger before continuing. Remember that SketchUp doesn't do a good job when creating curved surfaces at small scales.

6. Using the Scale tool, enlarge the model 10 times its size (Figure 6–19).

I decided to enlarge the model 10 times its size since it's a nice round number. But you can also enlarge it 20 or 30 times. After you create the arced surface, the model will need to be scaled back to its original size.

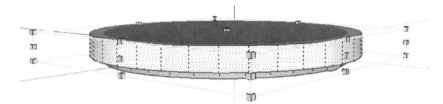

Figure 6–19. Enlarging the model 10 times its size

7. Now select the Follow Me tool, and select the arced surface you created (Figure 6–18b).

8. Follow a path around the model, ending at the start.

Your model should look like Figure 6–20.

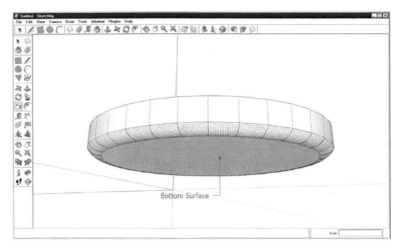

Figure 6–20. Creating an arced surface around the model

9.   Select the entire model, and scale it to .1 its size.

     You enlarged the model by 10 in step 6. Now you are scaling it back to its original size.

10. Extrude the top inner surface by 2mm, and then draw a triangle attaching the two top surfaces (Figure 6–21a). Using the Follow Me tool, select the triangle, and follow a path around the circle (Figure 6–21b).

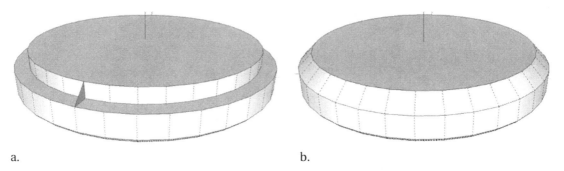

a.   b.

Figure 6–21. Blending the top surface

Modeling the Handles

Next you'll create the handles on each side of the sundial to which the watchbands attach. To assist you in the process of designing the handles, you will be using groups and components. Do you remember what the difference is between a group and component?

    Groups are great for combining a model or models into one for such tasks as copying and moving. Components also allow for easy moving and copying, but the added benefit is that making one change within a component will change all the other copies of it as well. This avoids the tedious task of having to edit each copy of the model separately.

1. Select the entire model, and right-click the model. From the drop-down menu, select Make Group.

   Now when you select the model, it will be surrounded by a blue box (Figure 6–22a). To edit the model, right-click the model, and select Edit Group, or double-click the model. In edit mode, the model will be surrounded by a dashed box (Figure 6–22b).

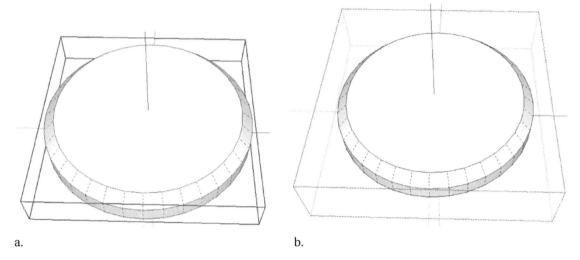

a.   b.

Figure 6–22. Grouped model and model ready for editing within a group

Right now you won't be doing any editing work within the model, so if you are in edit mode, click anywhere outside the dashed box, and automatically you will exit editing mode.

2. To construct the handles of the model first, place the model in top view. Also rotate the model so that one of the hidden lines that defines the model is aligned with the green or red axis.

3. Using the Line tool, draw a 3mm line out from one end of the model and 10mm out on both sides (Figure 6–23a).

   You will use these lines to assist you in drawing the handles of the model. The handles will be 20mm apart from each other.

4. Draw lines on both sides into the model (Figure 6–23b).

   Make sure to zoom in to see whether they are connected to the model. Broken lines in the model will not create surfaces that you can extrude.

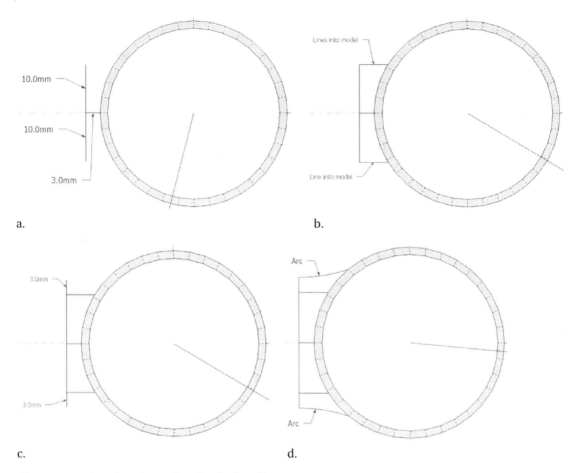

a. b.

c. d.

Figure 6–23. Creating the surface for the handle

5. Now create 3mm lines extending outward from both ends (Figure 6–23c).

6. Create arcs attached to the model (Figure 6–23d).

    All you need to do is close off the lines so that you have a solid surface to extrude.

7. Using the Line tool, create lines along the curve to complete the surface of the handle (Figure 6–24a). Now orbit to view the side of the model to see the two surfaces just created (Figure 6–24b).

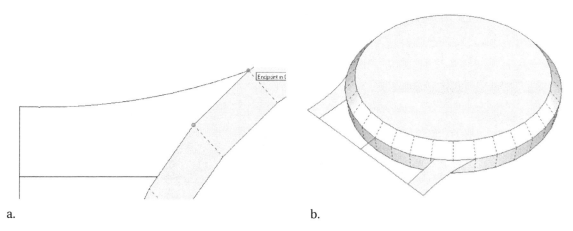

a.   b.

Figure 6–24. Surfaces created for the sundial handles

    8.   Using the Push/Pull tool, extrude the surface to the bottom part of the model (Figure 6–25a).

    9.   Then select the part of the model you just created, right-click, and select Make Component (Figure 6–25b).

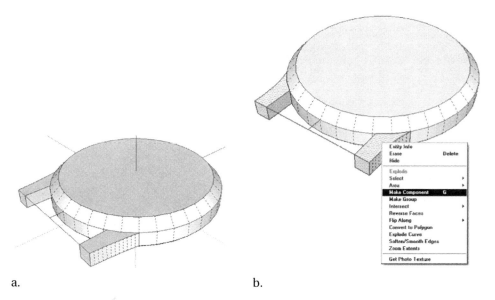

a.   b.

Figure 6–25. Converting the handle into a component

The Create Component dialog box appears. Enter **handle** for the component name, and select Create on the bottom right of the dialog box (Figure 6–26). Automatically the model will be surrounded by a blue box similar to the one created with Make Group.

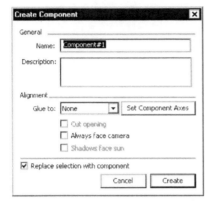

Figure 6–26. Create Component dialog box

Now you will be copying the handle component and placing it on the opposite end of the model.

10. Select the Rotate tool, and hit Ctrl on your keyboard.

   On the cursor, a + sign will appear indicating that you can now rotate the model and at the same time create a copy. Select and rotate the model (Figure 6–27a).

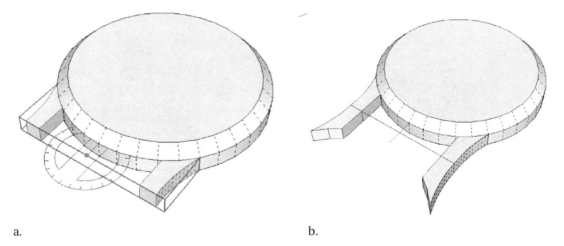

a.   b.

Figure 6–27. Copying and rotating the handle

If your model starts to look like a *Star Trek* space ship, you are probably on the right track (Figure 6–27b). Now that you have created a copy using the Move tool, drag it and attach it to the opposite end of the model. If you take a closer look at the model, you left behind a few extra lines, and you did not add the holes for the watchband to connect too. Double-click to access the component and delete the extra lines. To add the watchband holes, you will need to adjust the design of the handles slightly.

11. Extrude the handle from the backside by 2mm, and raise it to the height of the model (Figure 6–28a).

   I extended the backside of the handle to accommodate the size of the watch band.

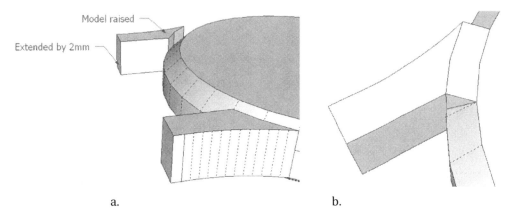

a.   b.

Figure 6–28. Extended handle

   All you need to do now is attach the hanging surfaces to the model. Using the Line tool, create lines connecting the surfaces to the rest of the model (Figure 6–28b).

   Now you will add the watchband holes so you can wear the sundial on your hand after you have 3D printed the model.

12. Create a guideline 1.5mm from the bottom and the side of the handle. At the intersection of the lines, create a 1mm diameter circle. The depth of the hole is 1mm (Figure 6–29).

   Repeat the process again on the opposite end of the watch handle.

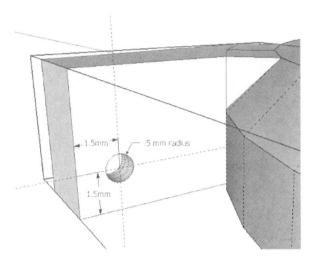

Figure 6–29. *Creating the whole for the sundial bands*

Since the changes were made within a component, the same changes will appear in the other component. That was great! You just cut the modeling time in half. This was just a small example of how components can help you save time during modeling. Wherever you see a duplicate copy of a part in your model, create a components of it so you won't have to waste the time having to reconstruct parts of it.

Placing the Dials

The next phase of the modeling process involves the design of the gnomon and the bullets defining the digits of the sundial.

1. On the surface of the sundial, create guidelines through the middle and one perpendicular line ending in the center of the model (Figure 6–30a).

   Double-click the model's surface to access the top surface of the model.

2. Using the Offset tool, create an offset 2mm inward (Figure 6–30b).

   Then exit editing mode. The offset will act as a guide when placing the hour markers on the sundial. Right off the top, you know that the location for 6 a.m., 6 p.m., and 12 noon will be at -90, 0, and 90 degrees. Using the Circle tool, create 1mm diameter circles at these locations (Figure 6–30c).

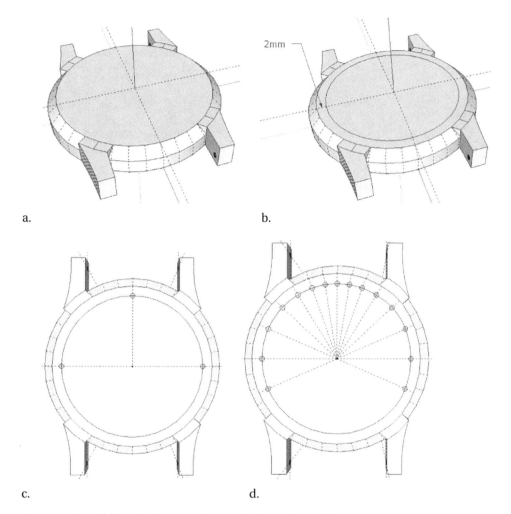

a.

b.

2mm

c.

d.

Figure 6–30. Adding the 7 a.m. to 7 p.m. digits to the model

From Figure 6–15, you know the locations for the other digits in the model.

3. Using the Rotate tool, click in the center of the sundial and once more on the perpendicular guideline. Hit Ctrl on your keyboard, and type **9.367** (the 1 p.m. digit).

This will copy the guideline and place it at 9.367 degrees. Repeat the process for each of the other degree locations (Figure 6–30d). Draw 1mm diameter circles at the intersection of the guidelines and 2mm offset from step 2. Next you will add the gnomon.

Designing the Gnomon

The gnomon will be at a 38-degree angle from the surface of the sundial and in the same direction as 12 noon. To make things easier, hide all the guidelines created except for the horizontal and perpendicular guidelines.

4. In the center of the sundial, create a rectangle 2mm wide and 12mm long (Figure 6–31a).

    By applying Pythagorean theorem, you can easily figure out the height of your block, which is 9.4mm.

5. Extrude the rectangle by 9.4mm.

    Draw a diagonal line from the top corner to the bottom corner (Figure 6–31b). Then extrude the top surface creating the triangle (Figure 6–31c). To get rid of excess material, I also took out part of the triangle (Figure 6–31d). All you need is the top part of the triangle.

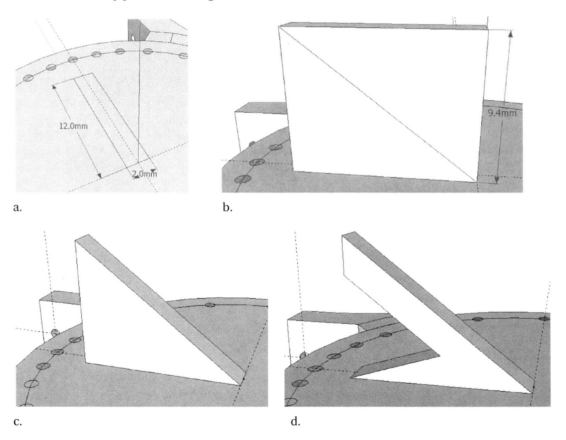

a.

b.

c.

d.

Figure 6–31. Creating the gnomon

Adding Text

Using the Line tool, draw a cross below the gnomon. It can be however you like; I have created a simple cross to indicate the direction of north, south, west, and east. To place 3D text, select the 3D Text tool. It is located in the Large Toolset and is represented by the A icon. The Place 3D Text dialog box will then open. Type **N** into the dialog box (Figure 6–32).

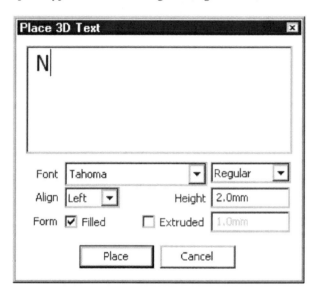

Figure 6–32. Place 3D Text dialog box

Within the 3D Text dialog box, you can adjust the font style, alignment, and size of your text. For the text height, type **2mm**, and deselect the Extruded check box. Selecting the Filled check box will fill the character instead of leaving it hollow. After you have made the changes, click Place. Attached to the cursor will be N. Click the surface of the model to place the text. You do not have to rotate the text. The text automatically rotates to the surface you are applying it to as the cursor approaches the surface (Figure 6–33a). Repeat the same process for the other characters (Figure 6–33b).

When first applying the text, it might come out bigger or smaller than you need. The text size can be adjusted using the Scale tool. To edit the individual characters, right-click each character, and select Edit Component from the drop-down menu. Now you can change each individual character.

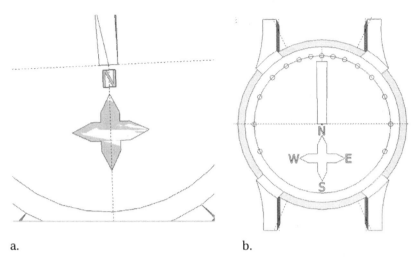

a.   b.

Figure 6–33. Name plate with changes to individual characters

To avoid the loss of any part of the model, let's create another group for the models on top of the surface. Select all the surfaces and parts, right-click, and from the drop-down menu select Make Component (Figure 6–34a). The model is now divided into three separate sections. You can see this in the Outliner (Figure 6–34b). There are two handles, the Bottom and Top groups. Within the Top group, there are the N, S, E, and W components. If you were to change the design of the handle and base of the sundial, you wouldn't have to worry about affecting the top surface.

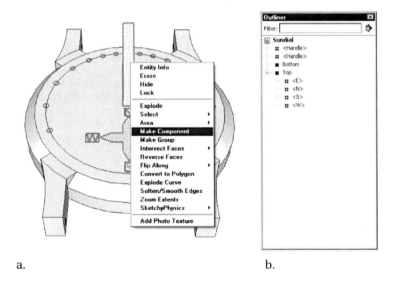

a.   b.

Figure 6–34. Sundial Outliner view

At this stage in the design process, I prefer saving the model under a different file name. Save the file as `Sundial_final` in case something goes wrong. It's good practice in case SketchUp crashes, because at least you will have a backup. Double-click the bottom group to access it, and delete the guide circle that you created, indicated by the blue line (Figure 6–35).

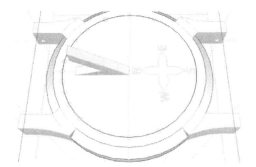

Figure 6–35. Deleting the extra circle acting as a guide

Next you want to explode all the groups and components in the model. Select each group and component in the model, and right-click them. From the drop-down menu, select Explode (Figure 6–36a).

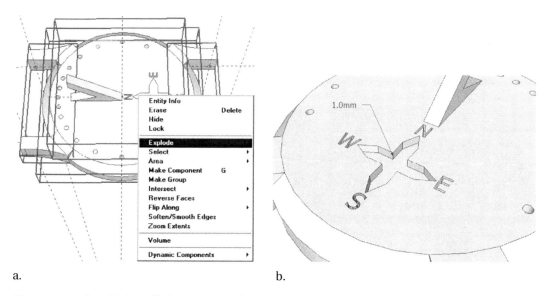

a.   b.

Figure 6–36. Combining all the groups and components

Explode takes out all groups and components in the modeling window and combines the entire model into one. Extrude all the circles, characters, and cross 2mm into the model (Figure 6–36b). Now that the model has been completed, the next step is to test the design.

Testing Your Sundial with Shadows

Here you will be using the Shadow Settings dialog box to test the functionality of the sundial. You can open the dialog box by selecting Window ➤ Shadows from the menu bar.

The great thing about the Shadow Settings dialog box is that everything is built-in. With a simple click, you can create any type of shadow effect. Shadows within SketchUp are not displayed automatically; therefore, to display shadows, you will need to select Show/Hide Shadows within the Shadows Settings dialog box or select View ➤ Shadows. You can apply shadows based on the time and day of the year. Drag the slider in the dialog box to adjust the time and date to your current time. The light and dark sliders are used for controlling the contrast of the model. Select the "Use sun for shading" check box, and you can hide/unhide the light and dark contrasts. At the bottom of the dialog box, you will find a selection of check boxes: On faces, On ground, and from edges. "On faces" creates a shadow on a surface of the model. "On ground" creates a shadow of the entire model projected onto the ground. "From edges" casts shadows from edges that are not part of a face (Figure 6–37).

Figure 6–37. *Shadow Settings dialog box*

Rotate the entire model, making sure north is pointing along the green line. The solid green line points north, and the solid red line points east. Set the time to 12 noon, and then select Show/Hide Shadows. As you can see from Figure 6–38, the 12 noon shadow has been cast. It looks like our sundial is working. Adjust the time within the Shadow Settings dialog box, and notice the shadow cast change for each time period.

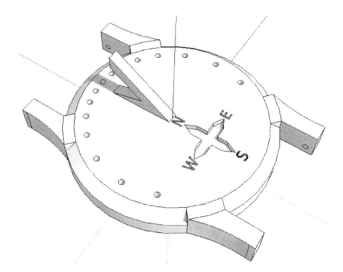

Figure 6–38. Sundial watch ready for upload

Double-Checking

At this stage, you are almost ready to upload the model for 3D printing. One last thing you will need to do is double-check the model for errors. Apply the five rules you learned earlier in this chapter to check your model. Since you combined multiple parts of model into one, there is a high chance that there are some internal surfaces to the model you don't need. Figure 6–39a shows internal surfaces left behind as a result of curved surfaces you created and multiple extrusions. Internal surfaces are also located at the intersection of the base and handle (Figure 6–39b). Select and delete these surfaces.

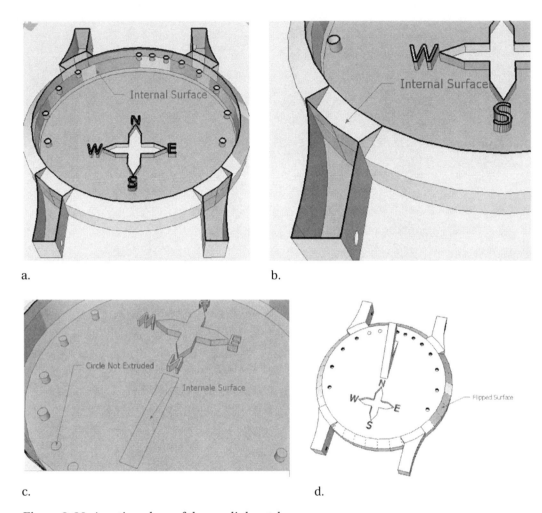

a.

b.

c.

d.

Figure 6–39. *A section plane of the sundial watch*

When applying a section plane from the bottom of the model, there are internal surfaces left behind when creating the gnomon. And couple of the circles in the model were not extruded either (Figure 6–39c). Figure 6–39d shows flipped surfaces on the outside of the model. These are a few things you will need to look out for and fix before uploading your model to Shapeways.

Uploading Your Design

Once you have fixed all the errors in the model, the next step is to export the file as a Collada file and upload it to Shapeways (Figure 6–40a).

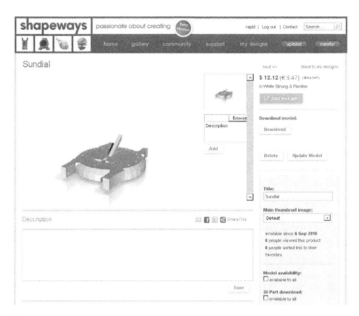

a.

b.

Figure 6–40. Sundial watch on Shapeways

The model only costs $12.12 to 3D print. After receiving the 3D print, I noticed there were some rough edges on the side of the model (Figure 6–40b). Increasing the number of sides of the circle used when creating the base will produce a smoother print on the edges. The characters on the sundial are visible and have come out quite well. You can now add a band and use it outside during the day to tell time.

Summary

What an exciting chapter! You started the chapter by designing a chess pawn where you learned about developing models with curved surfaces. Then you switched gears and created a sundial, utilizing groups and components. You also looked at shadows and how they can be applied to test the sundial. The next chapter is also very exciting. You'll learn to use a photograph to construct a 3D model for printing.

■ ■ ■

Modeling with Photographs

Welcome to Chapter 7! You have come a long way in the book, and I hope you are enjoying the experience. If there is one built-in feature that stands out the most in Google SketchUp, it is Match Photo. With Match Photo, you have the ability to create models from photographs. With a photograph, time is not spent brainstorming, sketching drafts, or collecting measurements—instead, you can go straight into developing the 3D model.

In this chapter, you'll skip the steps of creating sketches and brainstorming ideas and jump straight into modeling. You will start by becoming acquainted with all the options in Match Photo, and then you will construct part of a table to get your fingers warmed up. Once you understand the basics, you will dive in and construct a model of a house using Match Photo and prepare it for upload and 3D printing on Shapeways. Along the way, you will learn how to calibrate SketchUp's camera position and how inferencing can assist you in designing a model. By the end of the chapter, you will have learned how to add images to interior and curved surfaces of a model.

Creating a Simple 3D Model with Match Photo

Have you ever wanted to see a photograph you have taken developed into a 3D model? It's an exciting experience. In this section, you will model part of a dining room table using Match Photo so you can get a feel of how the different options work, and then in the next section, you'll construct a house that you will then 3D print using Shapeways. But before you jump in and start modeling the dining room table, you should understand the overall process:

1. You will need a photograph of the object to model. Google recommends the photograph be at a 45-degree angle from the corner of the structure. Make sure that the corner of the structure appears in the middle of the photograph. This provides good visibility of both sides of the model for tracing. If the corner isn't exactly in the middle of the photograph, no worries—you'll still be able to model the photograph.

2. Next you will match the photo in SketchUp by adjusting SketchUp's camera position and focal length so that it matches the camera settings that you took the photograph with.

3. After alignment, you are ready to trace the model using the Line tool. Remember that you will need to trace the model starting from the center of the axis, making sure every additional line you draw is attached to the previous line drawn. If you start drawing lines randomly, they will appear detached from the photograph. This will be clear once you start modeling.

Now that you have got a handle on what needs to be done and what types of things you will need to look out for when developing models using photographs, you'll now apply these steps and design part of a dining room table in SketchUp.

Modeling the Dining Room Table

The Match Photo feature in SketchUp allows you to reconstruct a model in three dimensions easily without having to take measurements or draw sketches of your model beforehand. Figure 7–1 shows an image of a dining room table that you will be using to demonstrate how Match Photo can be used. This example is to familiarize you with all the options in Match Photo that you can use. You can follow along in the design of this model by constructing it in SketchUp or simply read this section to get familiar with the available features that are part of Match Photo. If you want to follow along with the steps in this section, you can download example files for this book from the book's catalog page on the Apress.com web site. Look on the catalog page for the Book Resources section, which you should find under the cover image. Click the Source Code link in that section to download the example files. Unzip the download file, and the image is located in the Chapter 7 folder titled Table.jpg.

Figure 7–1. Photograph used for Match Photo modeling

The first step in the modeling process is to add the image to SketchUp's modeling window. There are two ways to add an image in SketchUp. The first way is to select File ➤ Import, and browse to the Chapter 7 folder (Figure 7–2). The image used for this example is saved in the JPEG format, but you can use most standard image formats. In "Files of type," select JPEG, and select Use as New Matched Photo from the lower-right corner of the dialog box.

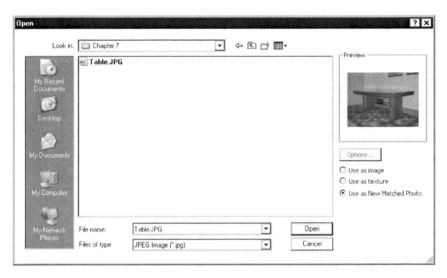

Figure 7–2. Importing an image to the modeling window

Click Open. SketchUp will automatically place the image into the modeling window as a new Match Photo image (Figure 7–3).

The second way of adding images to SketchUp is to select Camera from the menu bar and then select Match New Photo. Browse to the location of the saved image, and click Open.

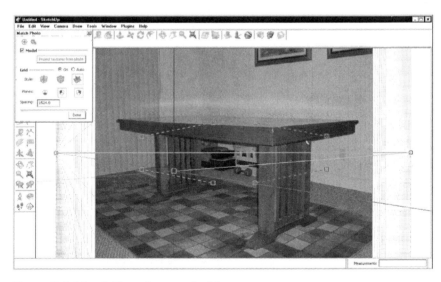

Figure 7–3. Match Photo imported table

After importing the image as a new Match Photo, you'll see an assortment of colored lines: the dashed green, the dashed red, the solid red, the solid green, the solid blue, and the solid yellow. The dashed lines represent vanishing point bars. The yellow solid line represents the horizon. Adjusting the horizon will move the vanishing point bars, and vice versa. All of this may look confusing at first, but don't worry. After rearranging the lines, the photograph will be ready for modeling.

Follow these steps:

1. Drag and place the red vanishing point bar grips along the top edge of the table.

   Repeat the same process for the second red vanishing point bar, and place it along another edge that is parallel to the first.

2. Line up the green vanishing point bar grips to an edge of the table that is perpendicular to the red vanishing point bars.

   Repeat the same process for the second green vanishing point bar, and place it along another edge that is parallel to the first green vanishing point bar.

3. Finally, click and hold the origin (yellow square); the cursor changes into a hand. Drag the origin, and align it perpendicular to the red and green vanishing point bars (Figure 7–4).

   Place the origin where all three axes (red, green, and blue) might intersect, and also line it up in parallel with the bottom edge of the table. A good place for the origin is at the front-bottom corner of the table.

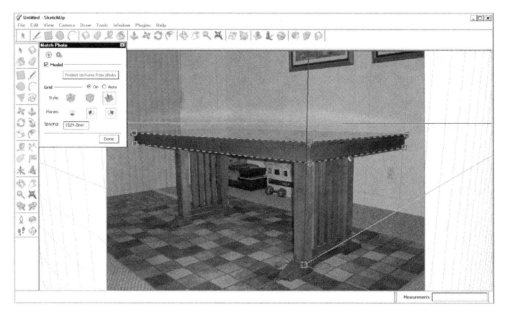

Figure 7–4. Match Photo imported table after alignment

But wait—let's take a closer look at the Match Photo dialog box to review some of its interesting features (Figure 7–5). The dialog box is divided into two parts: Model and Grid.

Figure 7–5. Match Photo dialog box

To hide or unhide the model, select the Model check box. To project the image of the table onto the model as a texture, select "Project textures from photo." Under the Grid check box, you can select from an assortment of styles. The leftmost style is for indoor photographs where all sides of a room meet. The middle style is for photographs taken from the top point down at an angle to the building or structure. The axis would lie at the top corner. The rightmost style is for photographs taken standing on the ground. For the table shown earlier, I have used the rightmost style.

Click the Red/Green button to toggle the red and green planes. Click the Red/Blue button to toggle the red and blue planes. Click the Green/Blue button to toggle the green and blue planes.

Changing the Spacing value sets the scale of the model. If this is your first time using Match Photo, I recommend you stay with the default settings. Once you have lined up everything, click Done in the Match Photo dialog box. You are then ready to trace the model. But before you start tracing, there are a few rules to remember:

- Trace the photograph starting from the origin.

- Trace the photograph in a path parallel to one of three planes. This can be along the red, green, or blue axis.

- Trace every line, making sure it starts at the origin or starts at the edge or end of the line you have already drawn while tracing.

These rules are really important to remember. They will save you the frustration of having to redraw lines and surfaces in your model.

Tracing the Table

Now that you have aligned the vanishing point bars, you are ready to trace.

1. Select the Line tool, and trace the edge of the dining room table leg starting from the origin.

    Zoom into the corners of the model to make sure that all the lines are connected. After tracing is complete, you will have an outline of the leg (Figure 7–6). The outline of the leg is a little difficult to see since the surface is transparent, but zoom in, and notice there is a lighter shade to the surface. Also, the surface is sitting in parallel with the blue plane. Since the surface is along one plane, it will be easy to extrude.

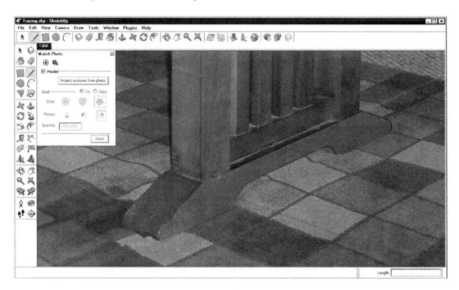

Figure 7–6. Tracing the table leg

2. Select the Orbit tool, and rotate around the trace.

    What you should see is the surface of the trace and not the complete trace in 3D form (Figure 7–7a). What we need to do now is extrude the surface to the width of the leg. On the upper-left corner of the screen is a tab called Table. The tab is created when you import an image with Match Photo into SketchUp's modeling window. The tab is named after the image. Click the Table tab, and the 3D model will align with the photograph as you continue modeling.

3. To extrude the surface, select the Push/Pull tool. Click the surface you just created, and extrude it to the width of the leg. Figure 7–7b shows what your model should now look like after you are done.

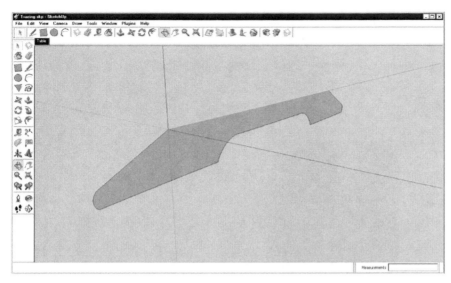

a.

b.

Figure 7–7. Tracing and extruding the surface of the model

Make sure when you are tracing any image to zoom in and out of the model to check whether the surfaces and lines are aligned with the photograph. Also, orbit the image frequently to search for lines that don't follow the path you had originally traced. Delete those lines, and try retracing them again. If this is your first time tracing an object in SketchUp, it might take couple of tries before you get a good feel of how tracing works.

4.   When the model is aligned with the photographic image, select the entire model, and right-click it. From the drop-down menu, select Project Photo (Figure 7–8a).

A dialog box will appear asking whether you want to trim partially visible faces (Figure 7–8b). Click Yes to apply the texture to faces shown in the image, or click No to apply texture to the entire face. Now the image will be projected onto the model's surface. Rotate the model to see what the projection looks like just on the model (Figure 7–8c).

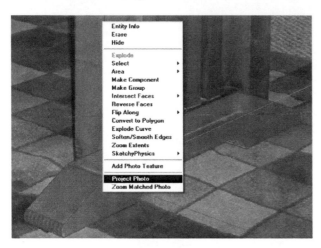

a.

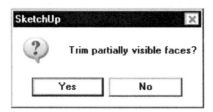

b.

c.

Figure 7–8. Projected image on model

You have just learned how to reconstruct a model in three dimensions with the Match Photo tool in SketchUp by modeling a small part of a table. You can apply a similar approach to other types of models. In the next section, you'll apply the same techniques learned here and use them to construct a house model; then you'll upload it for 3D printing on Shapeways.

Model a House for 3D Printing Using Match Photo

Now that you are all warmed up, it's time to get down and dirty and develop a model that you can 3D print using Shapeways. Before we continue, remember that there are three steps when designing models using Match Photo:

1. You will need a photograph of the object you will be modeling.

2. You will upload it to SketchUp and, using Match Photo, adjust its camera position and focal length.

3. Finally, you will trace the model using the Line, Arc, or Rectangle tool.

Importing the Photograph

In this section, you will be using a photograph of a house to construct the model. The photograph that you will be using is in the Chapter 7 folder and is titled House.jpg.

1. Start by importing the house image into SketchUp using Match Photo.

2. Save the import, and give it the name **House**.

Once the image is imported into SketchUp, you will see all the vanishing point bars, and the Match Photo dialog box will appear (Figure 7–9). This is similar to the one in Figure 7–3 when importing the table image.

Figure 7–9. Match Photo import of house photograph

151

In the next section, you will calibrate SketchUp's camera so that it matches the position and focal length of the camera that was originally used to take the picture before you start tracing the model.

Calibrating SketchUp's Camera

Before aligning all the vanishing point bars to the photograph, take a close look to see where in the photograph would be a good location to place them. You want to place the vanishing point bars along an edge that is parallel with the edge of the house.

1. In the modeling window, place the red vanishing point bars along the side of the house, making sure they are parallel to each other.

2. Place the green vanishing point bars along an edge in front of the house, with both in parallel to each other.

    Make sure the green and red vanishing point bars are perpendicular to each other. For better precision, extend the vanishing point bars along the entire length if possible.

3. Drag the origin, and place it in the front of the house at the left-bottom corner (Figure 7–10).

    The origin is the location from which you will start modeling the house, so make sure to align it so the axes fall along the edges of the house. In the imagery, there are some bushes that are blocking part of the patio area. Don't be too worried. There are other reference points within the photograph you can use to trace. In the Match Photo dialog box, stay with the default settings for now.

Figure 7–10. Photograph after alignment

4. After aligning the lines, click Done in the Match Photo dialog box.

Once the model is aligned correctly, you are ready to trace the side of the house.

Tracing the House Photograph

Start with the Line tool, and click the origin. Trace the edge of the house while making sure that each line you draw is connected to the line before it. Draw only the side surface of the house (Figure 7–11). You can also draw the surfaces that make up the front of the house, but tracing those would result in extra work.

Figure 7–11. *Side tracing of the house photograph*

Working with Inferences

While you are tracing the side of the house, take advantage of inferences in Google SketchUp. Inferences appear automatically while you are modeling to identify points and lines of interest. If you place the cursor on an inference point, it will display the inference. Each inference point is defined by a different color in SketchUp. These are On Edge, On Face, Midpoint, Endpoint, and Intersection. Figure 7–12 shows each type of inference point.

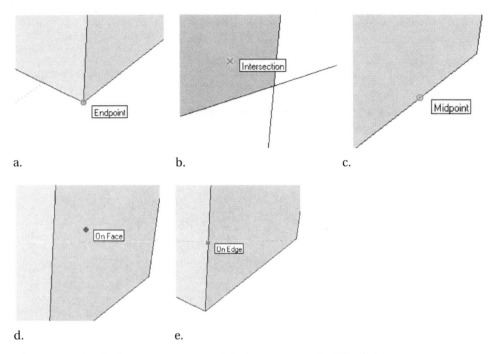

Figure 7–12. *Endpoint, Intersection, Midpoint, On Face, On Edge inferences*

Table 7–1 describes each of the inferences shown in Figure 7–12.

Table 7–1. *Types of Inference Points*

Name	Description
On Edge	A red point appears on the edge of a line.
On Face	A blue point appears on a face.
Endpoint	A green point appears at either end of a line.
Midpoint	A cyan point appears in the middle of the line.
Intersection	A black point appears at the intersection of a line.

Inference lines in SketchUp appear as solid colored lines or dashed colored lines. Figure 7–13 shows each of these lines.

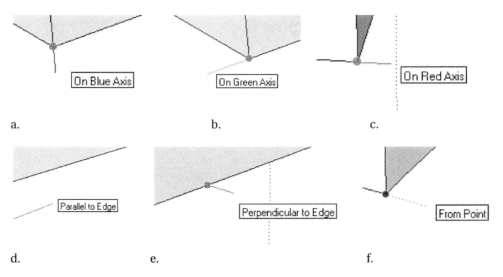

a. b. c.

d. e. f.

Figure 7–13. On Blue Axis, On Green Axis, On Red Axis, Parallel to Edge, Perpendicular to Edge, and From Point inference lines

Table 7–2 describes each type of inference line.

Table 7–2. Types of Inference Lines

Name	Description
One Blue Axis	Blue solid line aligned along the blue axis
On Red Axis	Red solid line aligned along the red axis
On Green Axis	Green solid line aligned along the green axis
From Point	A dotted line showing linear alignment from an endpoint
Perpendicular to Edge	A solid magenta line that is perpendicular to an edge
Parallel to Edge	A solid magenta line parallel to an edge

At times it can be difficult to see inference lines when there are multiple objects in your modeling window. To assist you, you can lock the inference line by holding down the Shift key. The inference will remain locked even if you move your cursor or point to another inference. Once an inference line is locked, it will appear in bold. You can also force an inference lock by holding down the arrow keys. The right arrow key locks the inference along the red axis. The left arrow key locks the inference along the green axis. The up and down arrow keys lock the inference along the blue axis.

Cleaning the Trace and Extruding the Surface

After adding the traces to the photograph, extra lines and surfaces are created. Select and delete the lines and surfaces that do not define the outer edges of the model. Also, delete any unnecessary lines to reduce the number of surfaces in the model. In Figure 7–11, there are ten surfaces in the model that make up the side of the house. By deleting all the extra lines and surfaces, you end up with only two surfaces (Figure 7–14). The model is much cleaner now.

Figure 7–14. Two surfaces after deleting the extra lines and surfaces

Use the Push/Pull tool, and extrude both surfaces to the opposite end of the house (Figure 7–15). As you are extruding the surface, zoom to the other end of the house, and make sure not to go beyond its length. It's difficult to see where the other end of the house is if you don't zoom in. As a reference point, choose a corner of the house within the photograph to extrude to.

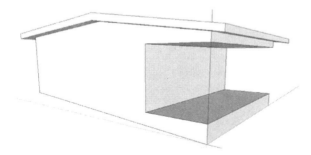

Figure 7–15. Extruding the surfaces in the house

156

If the roof appears detached while extruding from the rest of the house, then you might have drawn lines that project outside of a plane that one of the vanishing point bars is placed on. If you come across this situation, delete the roof, and try retracing it again. This time use inferences while drawing the roof. I drew 150mm lines tangent to the side of the house as guides to draw the roof (Figure 7–16). After tracing the roof, you can delete the extra lines.

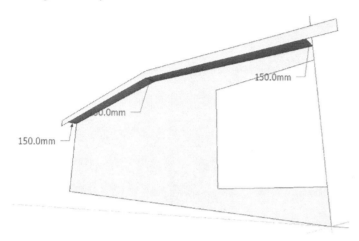

Figure 7–16. Drawing 150mm lines to trace the roof

Now rotate the model to the other side, and make sure you did not overextend the roof. Using the Tape Measure tool, measure the distances of the overhang (Figure 7–17). Make sure the overhang on both sides of the model is proportional.

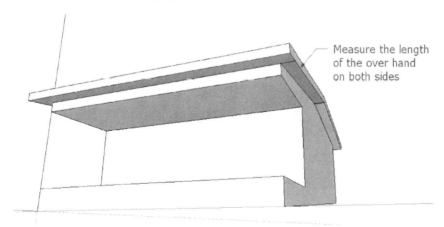

Figure 7–17. Extended overhang of the roof

Projecting the Photo

Select the House tab on the upper left of the modeling window, and align the model with the photograph. Right-click a surface of the model, and from the drop-down menu select Project Photo (Figure 7–18). Repeat the same steps for the other surfaces.

Figure 7–18. *Projecting the surface of the model*

Now with all the surfaces projected, you can rotate the model to view all its sides while keeping the photo texture in place (Figure 7–19).

Figure 7–19. *All surfaces projected*

In case the projection does not line up accurately with the model, try Fixed Pin mode to adjust the photo texture. From the drop-down menu, click Texture, and then click Position (Figure 7–20a). You are placed in Fixed Pin mode. In Fixed Pin mode, you can use the pins to skew, scale, shear, and distort the texture to fit your model (Figure 7–20b). Use the Move Pin to adjust the position of the texture. Use the Scale/Rotate Pin to scale and rotate the texture. Use the Scale/Shear Pin to resize, shear, or slant the texture. Use the Distort Pin to adjust the perspective correction of the texture.

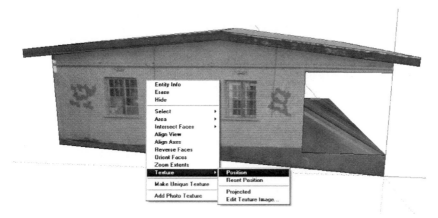

a.

b.

Figure 7–20. In Fixed Pin mode, you can skew, scale, shear, and distort the texture.

After adjusting the texture, right-click your mouse, and click Done. Now that the texture is in place, you can add detail to your model.

Adding Detail

Once you are happy with the alignment of the texture on your model, you can use the texture to trace the windows and doors. Select the Line tool, and trace one of the windows on the side of the house. Since both windows are the same, you can trace one and then copy and paste the other. Repeat the same process for the doors and windows in front of the house. Once all the traces are in place, use the Push/Pull tool to extrude the surface. Extruding the doors and windows will add definition to the model and will be visible after you 3D print the model. Once you are all done, your model should look similar to Figure 7–21.

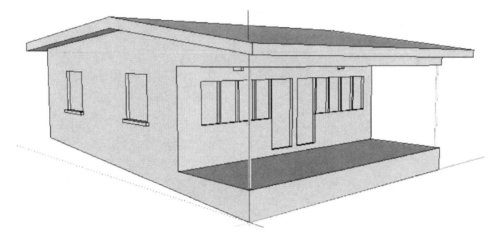

Figure 7–21. Modeling with windows and doors

Remember this is your chance to add as much detail you want to the model. In the next section, you will prepare the model for 3D printing on Shapeways.

3D Printing the House Model

Currently if you were to upload the model, it would be too costly and too big for the 3D printer to print. Using the Scale tool, you will reduce the size of the model so the width measures only 30mm.

1. Measure the width of the model using the Dimension tool.

    The model in Figure 7–22a measures 4135.5mm (l) × 3109.6mm (w).

2. Divide 30mm with the width of the model.

    The division of both numbers will give you a scaling factor that you can use to scale the model. Dividing 30mm by 3109.6mm is .0096.

3. Select the entire model, and using the Scale tool, scale the model by .0096.

After scaling the model, it measures 39.9mm (l) × 30.0mm (w) (Figure 7–22b).

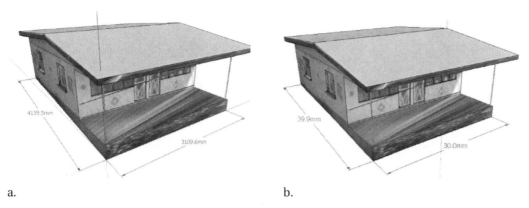

a.   b.

Figure 7–22. Before and after scaling the model

After scaling the modeling, remember to double-check your model for any errors. Refer to Chapters 4 and 6 where I discuss some of the errors you should look out for while modeling. After you're done double-checking, upload the model to Shapeways to see how much it costs.

■ **Note** Throughout your modeling adventure, it is common to come upon error messages. Don't be discouraged to see an error message. All that the message lets you know are some of the things to fix in your model. Right after uploading the house model, I got an error message (Figure 7–23). The message says "Only manifold objects can be printed." Shapeways sends most error messages to your e-mail address, so make sure to check your e-mail after every upload. In case you do come across this error, refer to Chapters 4 and 6 where I discuss how to make your model manifold.

Hi,

We just finished processing your product, House, and ran a few checks on it to make sure your product can be printed.

Unfortunately, we were not able to process and thus print your product. The error message received is the following "Only manifold objects can be printed".

If you have any questions about your product, please visit www.shapeways.com/mydesign under the header "my errors" or have a look at our relevant tutorial or video. Alternatively, you can send us an email service@shapeways.com.

Kind regards,

The Shapeways Service Team
Ralph, Maartje, Tyce and Kevin
service@shapeways.com

Shapeways
Passionate about creating
www.shapeways.com

Figure 7–23. Shapeways manifold e-mail error message

Pricing the Model

After fixing the error and uploading the model, you can see that the model costs $23.61 (Figure 7–24). If necessary, go back to Chapter 4 for a refresher on uploading your model for pricing. You can further reduce the costs of the model by scaling it, or you can get rid of some access material in the model. For such a small model, scaling it even further would not be beneficial. Instead, you will hollow out the model.

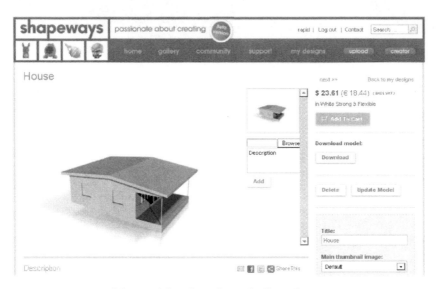

Figure 7–24. Price of the model without being hollowed out

To make the house model hollow, follow these steps:

1. Rotate the model to view the bottom.

2. Use the Offset tool to create a 2mm offset on the bottom surface.

    The depth of the patio area is different from the rest of the house, so you will need to create a line to divide the bottom surface of the model (Figure 7–25a).

3. Use the Push/Pull tool to extrude the surface into the model.

Do not extrude the model too close to the roof. Leave yourself a 2mm gap. Shapeways recommends a 2mm minimum wall thickness. Anything smaller will affect the integrity of the model. The model in Figure 7–25b has a depth of .3mm under the patio and 10.6mm under the rest of the model (Figure 7–25b).

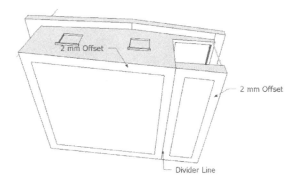

a.

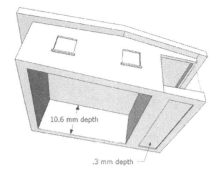

b.

Figure 7–25. *Bottom view of model*

After uploading, you can see there is a drastic drop in the price of the model (Figure 7–26). From a solid to a hollow design of the house, we saved $10.89—almost half the price of the initial model. As a designer, it is very important to take into account the model's overall volume. Small changes in volume can double or triple the price of the model. Remember as you are modeling to keep a close eye on areas in the model where you could reduce volume while at the same time staying within the limits of the material.

Figure 7–26. *The price before and after being hollowed*

Adding Width to the Roof

After a close inspection of the model, you'll notice the width of the roof measures only 1mm (Figure 7–27a). The 1mm overhang could chip off after 3D printing. You will need to increase the overhang width to 2mm. There are two approaches—either scaling the entire model larger to increase the overall width of the roof by 2mm or just increasing the size of the roof. By scaling the model larger, you increase the volume of the entire model and the cost as well. By just enlarging the width of the roof, you only increase the cost of the model by a small amount and slightly distort the look of the model. This is a choice that you will have to make. I prefer extending the width of the roof by 1mm since it is less costly and the model still looks reasonably good.

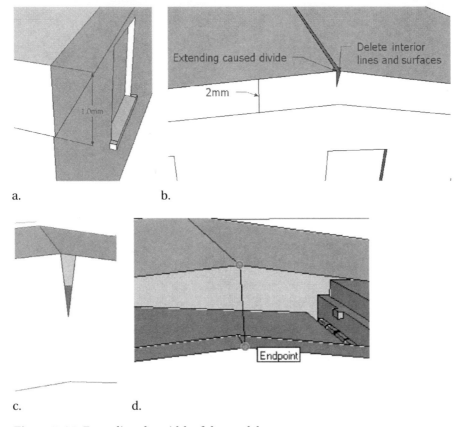

a.   b.

c.   d.

Figure 7–27. Extending the width of the model

To extend the roof, follow these steps:

1. Use the Push/Pull tool, and extrude the top surface of the roof by 1mm.

    The extrusion will cause a divide in the roof (Figure 7–27b). Zoom in, and delete the internal surfaces and lines that make up the divide.

2. Use the Line tool, and connect the two halves of the model.

To get rid of the triangles, delete the lines defining the triangle (Figure 7–27c).

3. To complete the roof, draw a line from top to bottom, completing the two halves of the roof (Figure 7–27d). Complete the same process on the other side of the roof as well. Since you have raised the roof by 1mm, you also can increase the depth of the inside by 1mm. The price for the model after upload is $14.07, an increase of $1.37—not bad at all!

Adding Images to Curved and Interior Surfaces

Earlier in this chapter you learned how to use Match Photo to construct a model from an image. You also saw how an image can be projected onto the surface of a table and house model. What if you wanted to trace an image within the interiors of a model like a cabinet, archway, or door in a house? You can draw them separately and place them in the house. Or in SketchUp, you can easily place an image within the interior and trace it. In addition, you can add an image on to a curved surface and then trace it. Using images to model saves time. Images allow you to develop detailed models without drawing sketches. Next I will show you how you can add an image to the interior and curved parts of a model.

Adding an Image to an Interior Surface

Adding an image to a interior surface in SketchUp is very easy.

1. Select File ➤ Import.

2. In the dialog box, select an image to import (Figure 7–28), and on the bottom right select Use as Texture.

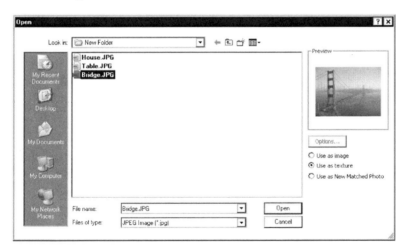

Figure 7–28. Importing an image to a flat surface in SketchUp

3. Click Open.

Notice that the image is attached to your cursor. Move the cursor to the wall on which you want to place the image. Figure 7–29 shows an interior wall constructed to place the image.

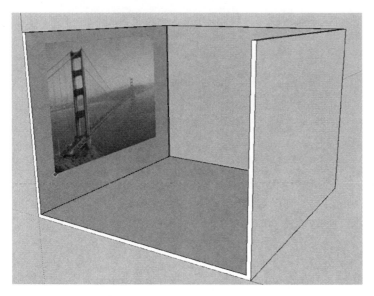

Figure 7–29. Attaching an image to the inside of a model

4. Click the bottom-left corner of the interior surface. Then click the upper-right corner of the surface.

As you move the cursor to the upper-right corner, the image will enlarge to fit the entire surface (Figure 7–30). If you want to add images to additional images, repeat steps 1–4.

Figure 7–30. Image adjusts to cover the entire surface of the wall

Adding Images to Curved Surfaces

So far you have placed an image on a flat interior surface. In this section, you will learn how to place an image on a curved surface. Yes, it can be done—but it requires a few extra steps. For this example, you will be using a cylinder as the curved surface (Figure 7–31).

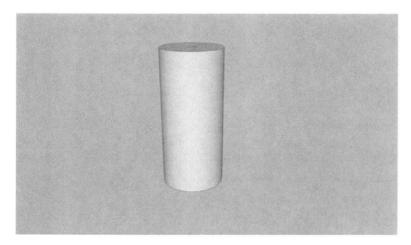

Figure 7–31. The cylinder used for texturing

1. In front of the cylinder draw a flat surface with the same height and diameter as the cylinder.

2. Select File ➤ Import, and select an image to import.

   Remember to click Use as Texture and then select Open.

3. Attach the image to the flat surface created in step 1.

   You will need to click the lower-left corner and click again in the upper-right corner of the surface to attach the image. Once the texture is in place, double-check and see whether you need to adjust the dimensions of the texture. This might require that you right-click the texture, select Texture, and then select Position. Using the yellow pin, adjust the size of the image to fit the flat surface. Once you have everything in its proper place, your model should have the texture in front of it (Figure 7–32).

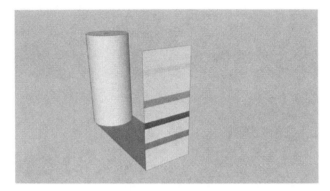

Figure 7–32. Texture attached to the flat surface

You will now have to project the texture onto the cylinder.

4. Right-click the flat surface, and select Texture ➤ Projected.

    This will stretch the image around the surface, giving the appearance of a complete image rather than being tiled.

5. Select the Paint Bucket tool, and then press and hold down the Alt key.

    The Paint Bucket tool will turn into a dropper. Click the texture, and the Paint Bucket tool will load the texture. You will then see your image displayed in the upper-left corner of the Materials menu.

6. Click the cylinder to apply the texture (see Figure 7–33).

Now you can delete the flat surface.

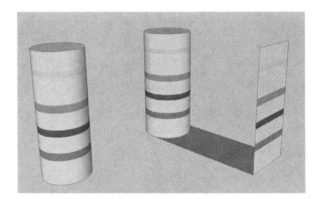

Figure 7–33. Textured image on a curved surface. Clicking any other cylindrical surface with the Paint Brush tool will allow you to infinitely texture identical images.

Using the Line tool, now you can trace in the details of the model.

Summary

This chapter was all about developing models with photographs. You started off with a brief overview of Match Photo by constructing part of a table using a photograph and then modeled a house using Match Photo for 3D printing on Shapeways. Along the way, you learned about calibrating SketchUp's camera and how inference can be used to assist in developing models. You saw how an image can be projected onto a model for tracing. We then concluded the chapter by seeing how to place an image within the interior and curved surfaces of a model. In the next chapter, you'll learn to apply symmetry in designing a model of an armored car.

CHAPTER 8

Working with Symmetry

So far in this book you have modeled a lighthouse, chess piece, sundial, and part of a table and house using the Match Photo feature in SketchUp. By going through each chapter and designing the models, you are starting to get use to the 3D modeling and printing process. If you have reached this stage in the book and still have not quite gotten your hands trained for modeling, fear not. It can take a couple of months to develop a modeling mind-set and get your hands familiar with the modeling process. Keep your head high, and take things one step at a time. Sooner or later, you will get it.

In this chapter, you'll step things up a notch and design symmetrical objects. Modeling symmetrical objects can be very exciting, and you will get a feel for it starting halfway in this chapter. The great thing about modeling symmetrical objects is that you need to model only half of them. The other half of the model can be easily duplicated. In this chapter, you'll apply many of the same techniques you have learned throughout the book to create a symmetrical model.

Symmetrical vs. Asymmetrical Design

In your daily life, you probably have encountered both symmetrical and asymmetrical objects. So, what are some examples, and what are the differences between both these types of models? Symmetrical objects when split in half look identical on both sides. Examples include cars, chairs, buildings, animals, and cups. There are endless symmetrical models you can find in the world. Symmetrical objects can also come in two formats. There are bilateral and radically symmetrical models. Objects that have bilateral symmetry are identical on both halves (Figure 8–1a). Objects that have radial symmetry can be divided into quadrants where each quadrant is similar to the first (Figure 8–1b). Radial symmetrical objects include umbrellas and staircases.

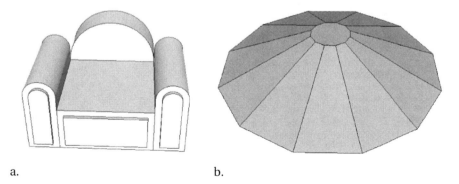

a.   b.

Figure 8–1. (a.) Bilaterally symmetric; (b.) radically symmetric

Objects that do not fall in the bilaterally or radically symmetric category are considered asymmetrical. Examples of asymmetrical objects include trees, houses, land, and rocks. Constructing these types of models requires more time since there are no sides to the object that are exact duplicates.

Designing Bilateral and Radically Symmetric Models

Let's go through two examples demonstrating the steps for constructing a bilaterally and radically symmetrical model. With the first example, you'll learn about bilateral symmetry by taking half a house and constructing a complete model. In the second example, you'll learn about radical symmetry by designing a staircase.

Modeling a Bilaterally Symmetrical Object

Creating a bilaterally symmetrical object can be broken down into six easy-to-follow steps:

1. The first step is to construct only half the model. The model in Figure 8–2 shows only half a house designed in SketchUp.

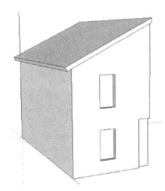

Figure 8–2. Half model of a house

2. Make the model into a component. To do so, highlight the entire model, right-click the model, and from the drop-down menu select Make Component.

3. Then create a copy of the component. Select the Move/Copy tool, and click a corner of the house. Press Ctrl once on your keyboard. The cursor will show a + sign alongside the cursor, indicating the model is ready to be copied. Drag the model, and click again to release it.

4. Now all you need to do is flip the copy of the model. You can use the Scale tool and invert the copy, or you can use Flip Along. Right-click the copy of the model, and from the drop-down menu select Flip Along. Select along the x-axis (red), y-axis (green), or z-axis (blue) (Figure 8–3).

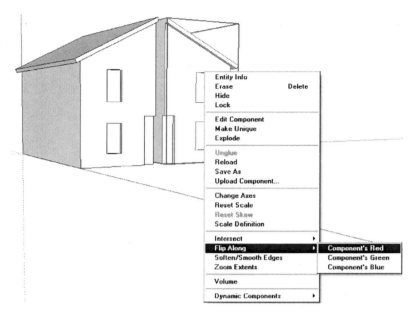

Figure 8–3. *Flipping your model along an axis*

5.   Using the Move/Copy tool, drag the two halves together to join the two components (Figure 8–4).

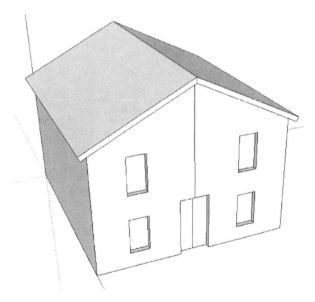

Figure 8–4. *Model after it had been flipped and put together*

6.   Since both halves are a component, changes in one will appear in the other. The model looks great, but there is something not quite right. Do you see the problem? Yes, there is a line through the middle of the model. To hide the line, double-click one component to enter it. Right-click the line, and from the drop-down menu select Hide. The lines are hidden, but the model is still in two halves. I don't think Shapeways will print two halves of a model. In the next section, you will see how to combine the halves for 3D printing on Shapeways.

Modeling a Radically Symmetric Object

The process to model a radically symmetric object is not different or difficult at all. Radially symmetric objects are made of quadrants, and each quadrant is the same as the first. An example of a radially symmetric object is a staircase. Let's see how a staircase can be modeled in SketchUp.

1.   The first step is to draw a polygon. Select the Circle tool, and click the center axis. Type **12s**, hit Enter on your keyboard, and then drag the mouse out to draw the 12-sided polygon. I chose 12 sides, but you can draw as many sides as you want. In this example, there will be 12 steps that make up the staircase.

2.   Draw a triangle using one of the edges of the polygon as a side of the triangle using the Line tool (Figure 8–5a).

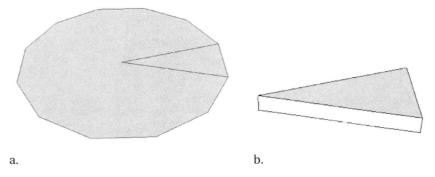

a.   b.

Figure 8–5. Using a polygon to create a triangle

3.   Delete all the other edges of the polygon, leaving behind only the triangle. You will use this triangle to create all the other steps in the staircase.

4.   Extrude the triangle using the Push/Pull tool to add width (Figure 8–5b).

5.   Select the triangle, and turn it into a component. As a component, you can make copies of each step easily. Also, in case you edit a step, all instances of the step will also apply those edits.

6.   Using the Rotate tool, click the vertex of the triangle, and hit Ctrl once on your keyboard. The cursor will show a + sign. Then click a second time on the left-outer corner of the triangle. Now rotate the triangle, making a copy that sits right next to the original (Figure 8–6a). Then type **12x**, and hit Enter. This will create 12 copies of the triangle alongside each other (Figure 8–6b).

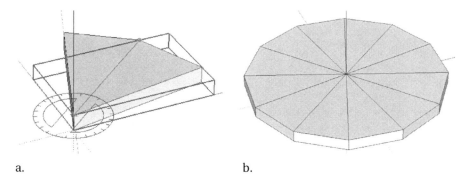

a.   b.

Figure 8–6. *Creating a radically symmetric model*

7. Select all copies of the triangle, and then create a copy using the Move/Copy tool on top of the first layer of the model. Remember to press Ctrl before making a copy. After the first copy, type **12x**, and hit Enter on your keyboard. This will create 12 copies one on top of the other (Figure 8–7a).

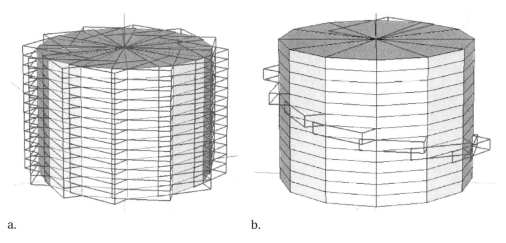

a.   b.

Figure 8–7. *Creating multiple copies of the model on top of each other*

8. Then select all the triangles from top to bottom diagonally while holding the Ctrl key (Figure 8–7b). Holding the Shift key on your keyboard, select the entire model, and then press Delete on the keyboard. This will leave behind a copy of the triangle you initially had selected while creating your radially symmetric staircase (Figure 8–8).

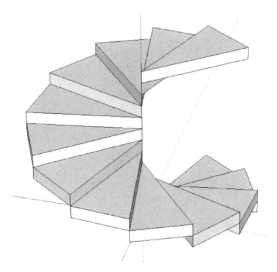

Figure 8–8. Radially symmetric staircase

In this example, there is no need to hide the edges of the triangle since each triangle is above the others on a different plane. Using a similar approach, you can create other types of models. There are umbrellas and starfish. Give this a try to see what you can make.

Designing a 3D Model from a Blueprint

This chapter is not over yet; it's time to get down to business. We have left the best model for last. You will construct an armored car in Google SketchUp, also known as the BA-64B armored car, which you then will upload for 3D printing on Shapeways. Unlike the models you constructed in the previous section, there are a few extra steps to consider when developing symmetrical models for 3D printing on Shapeways.

To design the BA-64B armored car, you will use a drawing to assist you along the way. You can download a copy of the drawing (Figure 8–9) from the book's catalog page on Apress.com.

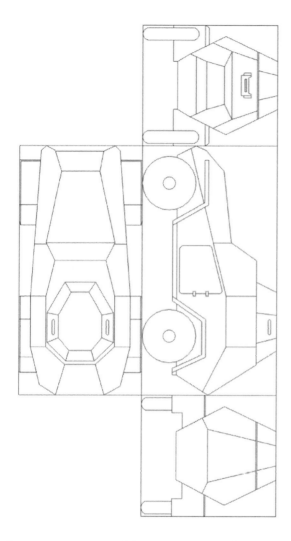

Figure 8–9. *Drawing of the BA-64B armored car*

Importing Four Views of the BA-64B Armored Car

To construct your model of the armored car, you need only four views (even fewer if you're a really good modeler). You will be using the side, bottom, front, and back views. You won't be tracing every detail in each view when constructing our model. You will use the views as a guide to construct parts of the model.

1. Using Microsoft Paint or any other image-editing software, crop the four individual views, and save them on your computer as separate image files in JPEG or PNG format. Name them A, B, C, and D for easy reference.

Make sure to crop just the image of the car and nothing more or less. This way it will be easier to align the images in SketchUp. If you are confused, don't worry. This will start to make sense when you upload the images into SketchUp.

2. In SketchUp, select File ➤ Import. The Import dialog box will appear; select the top view to import into the modeling window. Make sure to select "Use as image" in the dialog box. Then click Open to place the image in SketchUp.

3. Attached to the cursor will be the bottom view. Drag your cursor along the red or green axis, and then click the center axis. This will make sure the image is on the green and red axes. Drag your mouse outward to enlarge the bottom view, and click once more to lock the image in place (Figure 8–10a).

4. Now import the side view into your model. Click the Midpoint of the bottom view, and drag the cursor enlarging the image. Then click the opposite end of the bottom view (8–1-10b). If the image isn't perpendicular to the bottom view, use inferencing to align the image.

5. Import the back view, and attach it to the back of the side and bottom view. Click once to attach the image, and drag the cursor to enlarge the image. Click a second time to fix the image in place (Figure 8–10c). Repeat the same process for the front view. Once you are all done, your view should be aligned and ready for tracing (Figure 8–10d).

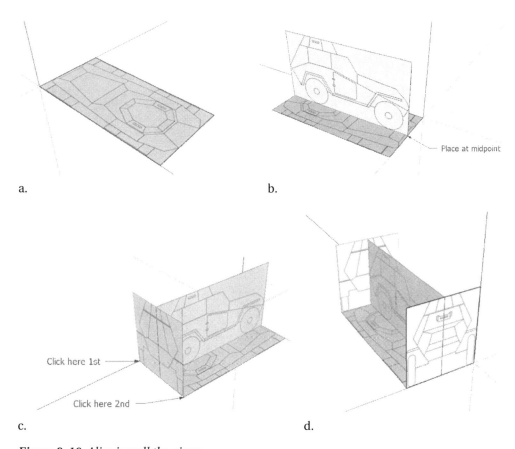

a.

b.

Place at midpoint

Click here 1st

Click here 2nd

c.

d.

Figure 8–10. Aligning all the views

■ **Note** If it's easier, you can enter the length of the photo after the first click. This will give you an exact match without having to drag and enlarge the image. Also for ease of use, rotate the views so all the surfaces are facing inward. When a trace appears on the opposite side of a view, the surface is flipped. This will save you the frustration of having to flip the views later.

6. Before continuing, save the model, and give it the file name BA-64B_Version_1. For a model this complicated, it's important to save multiple copies in case you need to go back to an original and make changes.

Tracing the Top and Front Views

Now that all your views are aligned in SketchUp, you are ready to trace your model. You will only be tracing the outer body of the armored car. Once you have the outer body of the model designed, you will then import the wheels and axle to complete the model by the end of the chapter.

1. To get started, trace the outline of the top view, as shown in Figure 8–11, using the Line tool. You have not traced the entire top view of the image. You only traced the flat surfaces of the model and the hood. You will create the other surfaces in a later section by looking at the side view.

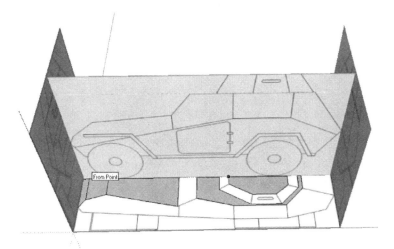

Figure 8–11. *Outline of top surface of model*

2. Now select the Move/Copy tool, copy each surface, and align it with the side view. The two surfaces at the end will be raised later (Figure 8–12). Notice the surface over the hood is not in alignment with the side image. You will need to align this to the side image.

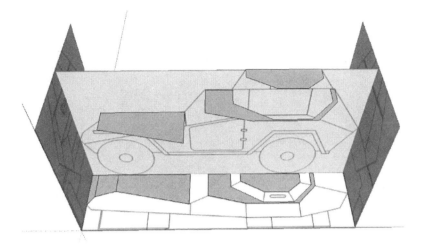

Figure 8–12. Aligning the traces with the side view

3.   Select the surface over the hood, and using the Rotate tool, move it so that it is in parallel with the slant of the hood. Now the surface of the hood is in alignment (Figure 8–13).

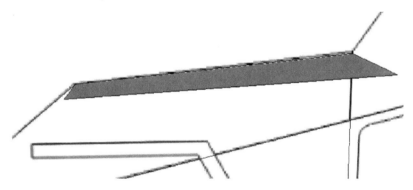

Figure 8–13. Aligning the surface with the slant of the hood

4.   You now raise a copy of the surface that attaches to the front of the hood using the Move/Copy tool and align it to the side view (Figure 8–14a). Then rotate it to match the slant of the side view (Figure 8–14b). If the surface is short compared to the side view, then, using the Line tool, draw in the extra surface (Figure 5-14c).

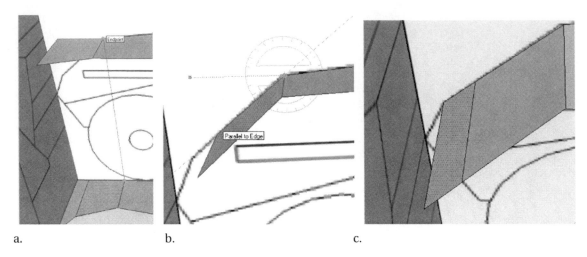

a. b. c.

Figure 8–14. Attaching the surface in front of the hood

5. The front of the model isn't complete yet. Using the Line tool, trace in the surface on top of the front grill. You will need to use the front and side views as reference to draw the lines. Figure 8–15 shows what the surface should look like after you are done.

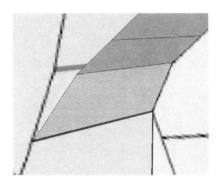

Figure 8–15. Tracing the surface above the front fender

6. To draw the grill, trace the front view of the model (Figure 8–16a). Using the Rotate tool, rotate the surface, and align it with the side view (Figure 8–16b). If the grill is short compared to the side view, extend it using the Line tool (Figure 8–16c).

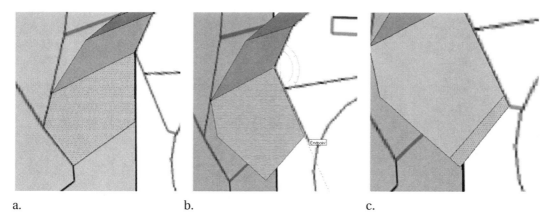

a.   b.   c.

Figure 8–16. *Tracing the front grill of the model*

Tracing the Side and Back Views

Tracing the side and back views is not difficult at all. You will be applying the same techniques used when you traced the top and front views in the previous section. Again, you will be using the Line, Move/Copy, and Rotate tools.

Tracing the Side

To trace the side view, follow these steps:

1. Use the Line tool, and trace the bottom half of the side view, as shown in Figure 8–17. Also draw in the horizontal lines defining the different surfaces of the model. You will be using these lines as a guide to draw in the other lines later.

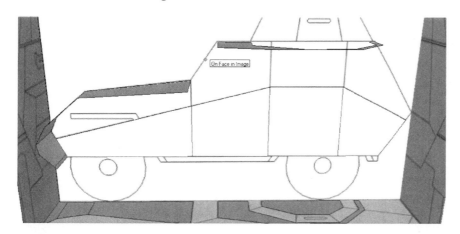

Figure 8–17. *Trace of the bottom half of the side view*

7. Now as before, using the Move tool, move the side surface, and align it with the front image.

▨ **Note** At this stage, you can also trace in the door. It is easier to do it now then later. Later I will be showing you how to add the door using a different method that requires more steps but is a great learning experience. To trace the door, select View ➤ Trace Style ➤ X-ray. The surfaces will appear transparent. Using the Pencil tool, you can trace the door.

---

8. Using the Rotate tool, rotate the side surface of the model. To move and rotate the surface, remember to select the entire surface and not just an edge, as shown in Figure 8–18.

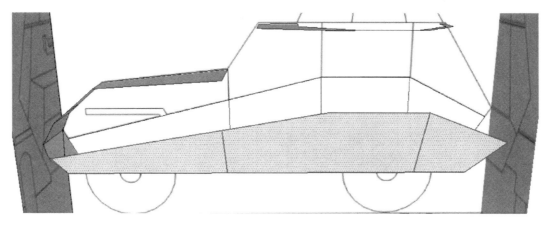

Figure 8–18. Rotating the side surface

As you did to trace the front view, you can take similar steps to trace in the back view using the Line tool. It is a tough model to draw in SketchUp, so take your time. If something doesn't feel natural while tracing, experiment to see what works best.

Constructing the Back Surface

To construct the back surface, follow these steps:

1. Trace the surface on the back of the armored car using the Line tool (Figure 8–19a). Then select the entire surface, and using the Move tool, align it with the edge of the side surface (Figure 8–19b).

2. Using the Rotate tool, rotate the surface so that it is in alignment with the side surface (Figure 8–19c).

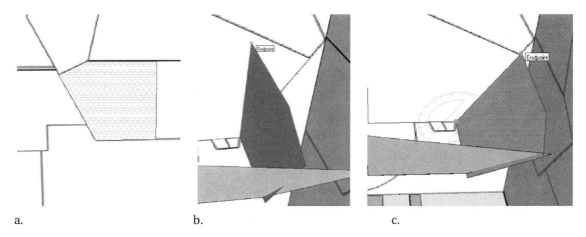

a.   b.   c.

Figure 8–19. Tracing the back surface of the model

> Looking at the model, closely notice that the surfaces are short and the back surface overlaps with the side surface.

3. Use the Line tool, and extend the back surface of the model (Figure 8–20a). Then delete and redraw the side surface of the model (Figure 8–20b).

> Notice SketchUp did not automatically fill the side with a single surface, so a diagonal line was created to complete the design. This is common actually when you design models of this type. The surface is not perfectly flat, and SketchUp uses two surfaces to connect the cap. This will be the case when working with surfaces that do not fall on the x, y, or z plane.

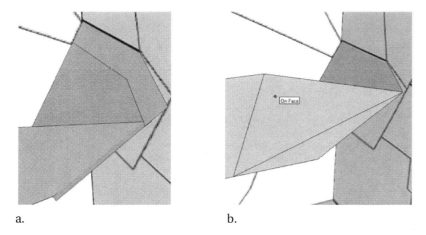

a.   b.

Figure 8–20. Extending the length of the back surface and redrawing the side surface to connect with the back surface

Filling In the Gaps

Now that you are done with creating the back surface, you are ready to add the missing gaps in the model. All the surfaces you have created so far were leading up to this stage. You will now be using those surfaces to assist you in filling in the gaps between those surface.

1. Use the Line tool, and draw lines from the side surface to the middle surface (Figure 8–21). Drawing these lines will automatically create a surface between the two. For surfaces that do not show a surface after adding the lines, try adding a diagonal line. Sometimes these surfaces consist of multiple lines.

■ **Note** Sometimes multiple lines are needed when an edge is made of more than one line. To solve the problem, delete the edge and redraw a single line. This may eliminate the need to redraw multiple lines.

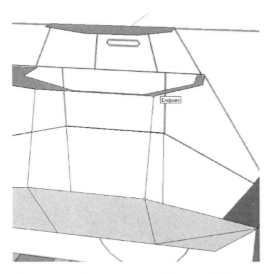

Figure 8–21. Connecting each layer with lines

As you connect each layer, the frame of the car will slowly come to form (Figure 8–22a). Have patience here; it might take a couple of tries to get everything right, so save a backup in case something goes wrong.

2. Hide the side view and rotate the model, and you will notice on the other side there is a hollow shell (Figure 8–22b). If there are any surfaces or lines within the shell, select and delete them. They may have been left behind in the process of constructing the model.

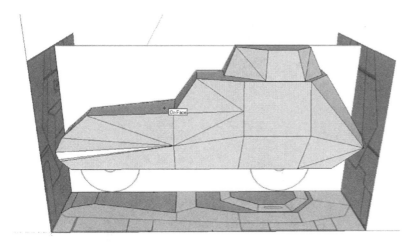

a.

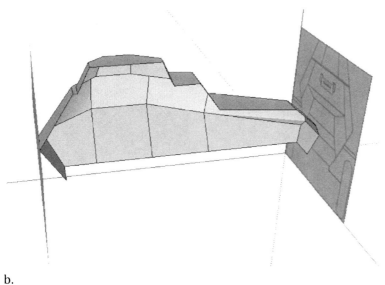

b.

Figure 8–22. Draw lines to fill in the gaps.

3. To get rid of the extra diagonal lines you created to draw in the surfaces, you
    can now hide them. Select the Eraser tool, and while holding down the Ctrl key,
    select each line in the model you want to hide.

4. Notice that some of the surfaces are facing into the model. Right-click the surface, and from the drop-down menu select Orient Faces. This will automatically orient all faces in the direction of the selected surface. This saves you time by not having to reverse each face/surface individually. Figure 8–23 shows the model after you are all done. The model is now much cleaner, and you can see all the surfaces that define the sides of the model.

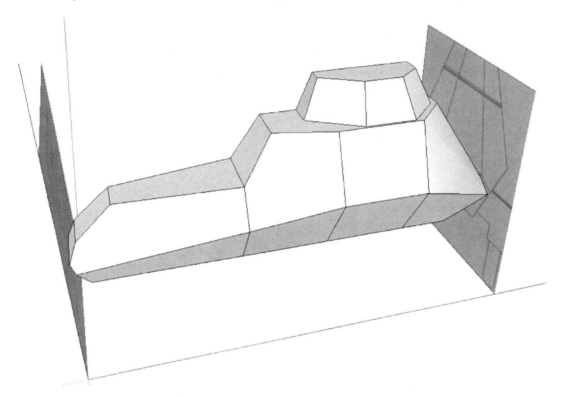

Figure 8–23. Model after the diagonal lines were hidden

Modeling the Fenders

You are almost done, but there are just a couple more things to add to the armored car before it will be ready for 3D printing. Next you'll add the fenders of the model.

1. Use the side image, and trace the fenders of the car (Figure 8–24a). After tracing the fender, hide the side image.

2. Using the Push/Pull tool, extrude the trace into the side surface of the model and out the other end (Figure 8–24b). Extrude it all the way to the edge of the front image.

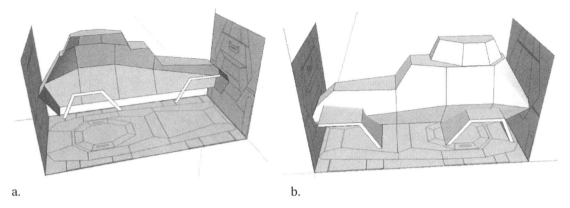

a.   b.

Figure 8–24. (a.) Trace the fenders; (b.) extrude the fenders out through the other end.

The fenders are not completely joined to the frame. You need to connect them before continuing.

3.   Use the Line tool, and trace the intersection of the fender with the body of the car (Figure 8–25). Also turn on hidden geometry. Since there are multiple surfaces that define the body of the car, it's important that you draw lines on those surfaces and not in a tangent.

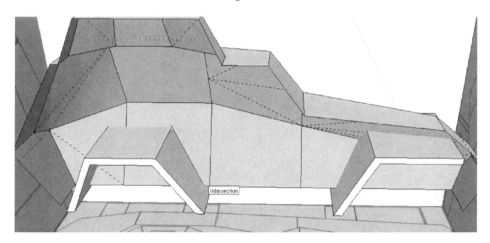

Figure 8–25. Drawing the lines to connect the fender to the chassis of the car

4.   After drawing in the edges, delete the fender surfaces within the model.

After you are all done, you should see an opening into the fender from the inside the model (Figure 8–26). This opening will be filled in with material when you send the model off for 3D printing to Shapeways. Make sure there is an opening, or you will be presented with an error during the upload process.

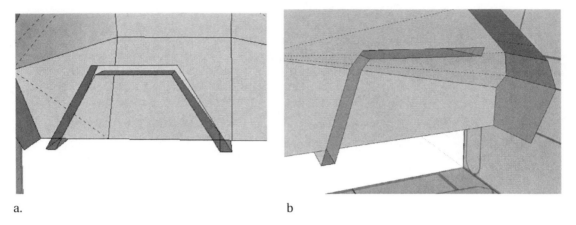

a.   b

Figure 8–26. Make sure to delete the extra material on the inside of the model.

Adding the Final Touches

Now that the fenders are in place, you will add some detail to the model such as the door and add windows in the armored car. There is actually a lot of additional detail that you could add to the model, but the process is long, so I will leave it up to you.

Modeling the Door

To model the door, follow these steps:

1. Trace the door and window of the model on the images as you did for the fender (Figure 8–27).

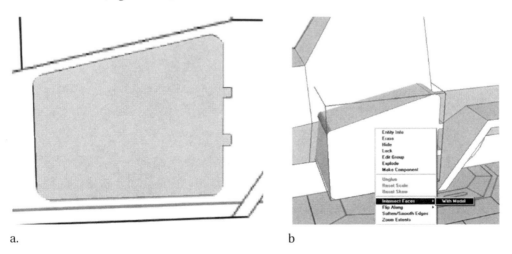

a.   b

Figure 8–27. Tracing the door of the model and extruding the door

2. After tracing the door of the armored car, select the entire trace of the door, and convert it into a group. Extrude the surface using the Push/Pull tool. Then right-click the group, and from the drop-down menu, select Intersect Faces ➤ With Model. At the intersection, a trace of the door will be created.

3. Select the grouped door, and hit Delete. What is left behind is just the trace. Notice the trace is slightly off from its original location. Since there is a slant in the surface on which the trace was created compared to the trace on the image, the door does not exactly line up with the model (Figure 8–28a).

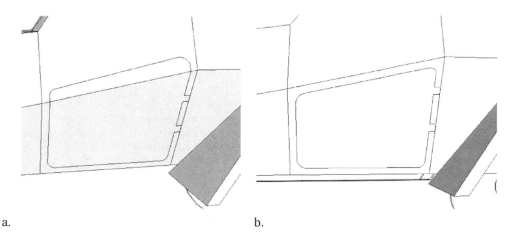

a.   b.

Figure 8–28. Trace projected on the surface of the body and adjusted to fit the surface

4. Using the Line tool, retrace the parts of the door in its correct location, and delete the lines in the wrong location. Once the door has been created, extrude it.

Modeling the Front Window

To draw the windows for the armored car, you can apply the same techniques as you did when drawing the door. But you can also trace the window and move it onto the surface of the car.

1. In Figure 8–29a, you have traced the front window of the car on the front view.

2. Use the Move tool, and move the trace onto the model. Use the Rotate tool to rotate the trace to match the slope of the front surface (Figure 8–29b). The window was extruded slightly to add definition (Figure 8–29c).

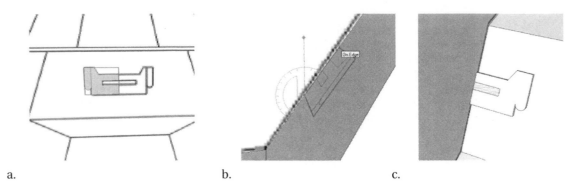

a.   b.   c.

Figure 8–29. *Modeling the front window of the armored car*

3. After adding all the details of the model, rotate the model, and make sure to delete all the internal remains, surfaces, and lines created as a result of adding the extra detail. Make sure to zoom into the model and delete any extra surfaces created. From the Figure 8–30, you can see when adding the window to the model it creates surfaces on the inner side of the model. Select and delete these surfaces.

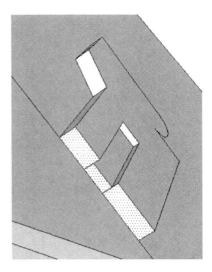

Figure 8–30. *Surfaces created on the inside of model*

Creating the Whole Model

You are about to see what the model looks like in its entire form, which is an exciting moment for a designer.

1. To complete the model, group the side of the model you just constructed, and then create a copy. Right-click the copy, and select Flip Along to flip its direction (Figure 8–31a). Drag the models together so they intersect and are in alignment with each other.

2. Select both models, and then right-click and select Explode (Figure 8–31b). This will combine both models into one. All you will be left with is a line down the middle of the model (Figure 8–31c).

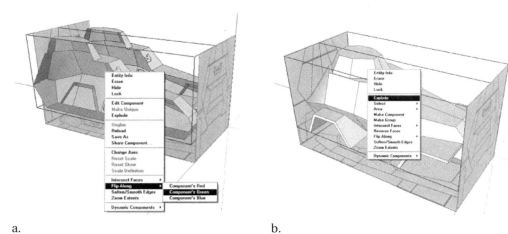

a.   b.

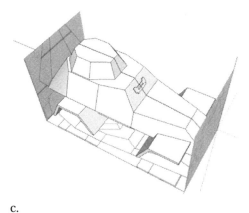

c.

Figure 8–31. *Flipping and aligning the two copies of the model*

3. Select the line down the middle, and delete it from your model. Now all that is left is to add the axle and wheel for the model.

■ **Note** If the line down the middle of the model is not perfectly straight and has a few gaps, try deleting the line and redrawing the surface combining the two halves.

Adding the Wheel and Axle

At this stage, you can go ahead and design your own custom wheel and axle to attach to the body of the armored car, or you can use the one I developed. You can download example files for this book from the book's catalog page on the Apress.com web site. Look on the catalog page for a section entitled Book Resources, which you should find under the cover image. Click the Source Code link in that section to download the example files. In the Chapter 8 folder there is a file titled axle and wheel.

1. Select File ➤ Import. Browse to the Chapter 8 folder, and insert the wheel and axle into your model. The wheel and axle are grouped separately. Align the wheel with the axle, as shown in Figure 8–32a.

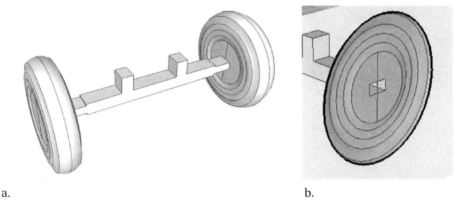

a.   b.

Figure 8–32. (a.) Axle and wheel aligned together; (b.) delete the surface created as a result of the intersection of the two groups.

2. Select both wheels and axle, right-click, and select Explode. The tires are now attached to the axle.

3. Create a section plane cut through one of the wheels; notice there are some internal surfaces generated as a result of the intersection (Figure 8–32b). Delete those surfaces on both sides.

4. Create a component of the wheel and axle combination, and then create a copy. Place one set under the front fender and another under the back fender. Align the tires with the bottom and side images.

5. After the axle and wheels are aligned with the body, right-click, and select Explode so they attach with the entire model. The model should look like Figure 8–33 after you are all done.

Figure 8–33. Axle and wheel attached to the chassis of the model

6. Select Window ➤ Section Plane. Create a section plane cut through the model, and delete the surfaces created by the intersection of the axle and body of the model (Figure 8–34).

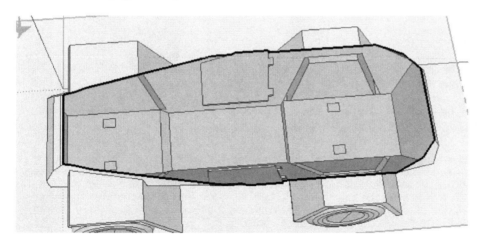

Figure 8–34. Section plane cut to delete the internal surface created between the axle and body

Uploading the Model for 3D Printing

All that is left to do now is scale the model and upload it for 3D printing on to Shapeways. Also, remember to double-check your model for any errors. Refer to Chapter 4 where I discuss what errors you should look out for when developing a model for 3D printing.

To scale the model, use the Tape Measure tool, and measure the length of the model. Divide 3 inches by the length. Remember the length has to be in inches. The division will give you a scaling factor. Enter the scaling factor to scale the model you want the model to measure 3 inches in length after scaling.

The model now measures 1.7 inches (H) × 3.0 inches (L) × 1.5 inches (W). Try to see whether you can upload the model at this state. I uploaded the model successfully without any problems into Shapeways, but the model costs $44.21. For a model this small, $44.21 is really expensive. Also, there is one thing that must be fixed. The width of the axle isn't 2mm. This might break the axle during printing. Shapeways recommends that the minimum wall thickness of any surface be at least 2 mm. Extend the axle so it measures 2mm in width, and hollow out the insides of the model to reduce the volume and as a result the cost of the model, as shown in Figure 8–35.

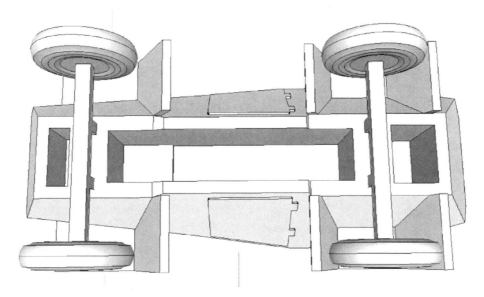

Figure 8–35. Extending the axle width to 2mm and hollowing out the model

After reuploading the model, you see that the model costs only $39.59—a small savings (Figure 8–36a). You can further reduce the cost by scaling the model even smaller, but further alterations would need to be made to the model that could compromise the overall design and increase the design time. As a 3D designer, these are some of the challenges you will face. So, make sure to plan before starting your design. From this point on, the design is a personal preference, so I'll leave rest of the alterations to the model up to you. Figure 8–36b shows the model after 3D printing.

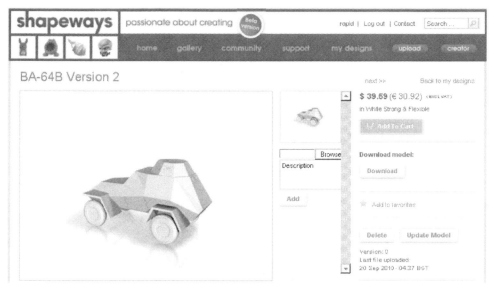

a.

b.

Figure 8–36. BA-64B armored care upload to Shapeways and after 3D printing

Summary

This chapter was quite an adventure. You started with an introduction to bilateral and radial symmetry. You learned how models can be designed with symmetry. Applying bilateral symmetry, you designed the BA-64B armored car. In the next chapter, you'll look at ways to share and sell the models you have designed in this book.

PART 3

Presenting, Sharing, and 3D Printing Alternatives

CHAPTER 9

■ ■ ■

Share with the World

Google SketchUp for 3D printing would not be complete without a look at the Shapeways Shop, 3D Warehouse, Google Earth, and Thingiverse.

- With the Shapeways Shop, you can open a store on the Shapeways web site to sell your 3D models.

- The 3D Warehouse is a SketchUp model repository. You can upload your own models and download models that others have uploaded into the 3D Warehouse.

- Use Google Earth to view satellite imagery of the world, view 3D buildings, and design your own buildings to be placed in Google Earth.

- Thingiverse is a web site for 3D modelers to upload and share their designs for 3D printing.

These are just a few things you can do once you have created your own models using SketchUp. We will go through each in this chapter and show you how easily you can upload and start sharing your designs.

Shapeways Shop

By now you probably are familiar with most if not all of the features available on Shapeways. You have learned about the different materials, have mastered the uploading process, and have gained knowledge about customizing your SketchUp models for upload to Shapeways for 3D printing. You also learned about the Creator and Co-Creator applications in Chapter 2. Throughout this process, you have gathered a collection of models from Chapters 4, 6, 7, and 8: a lighthouse, chess piece, sundial, table, house, and the BA-64B armored car. While designing each model, you learned something new about SketchUp and 3D modeling.

Now that you have built up a collection of models, it is time to sell them on Shapeways with your very own Shapeways Shop. The great thing about opening a Shapeways Shop is that the production, shipment, and customer service are handled by Shapeways. All you need to do is design and upload your models to your shop. You are paid on a monthly basis for every model that is sold, and all of the billing is done through PayPal. The models designed will belong to you. Before we continue, make sure you have a PayPal account. Also, access your account information on Shapeways, and double-check that all of it is correct. To create a Shapeways Shop, follow these steps:

1. Browse to the "my designs" page on Shapeways where you can view a gallery of all your uploaded models designed throughout this book (Figure 9–1).

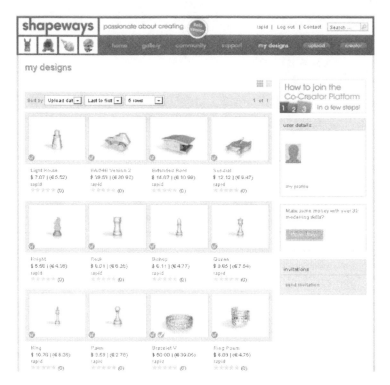

Figure 9–1. Shapeways "my designs" page

2. On the right side of the page is a green Open Shop button; click the button, and the profile page appears. Make sure the billing and shipping information is filled in and correct.

3. At the bottom of the page, click Create Shop. A menu will appear; enter a name for your shop, enter a URL, and agree to the terms and conditions. Then click the Save button (Figure 9–2). If you haven't filled in all the fields for the billing and shipping information, clicking Create Shop will do nothing.

3 Shop

Name:*   Rapid

Personal URL:*   http://www.shapeways.com/shops/   rapid

Agreement:*   ☑ I agree with the terms in the Terms & Conditions 🗔.
                         Fields marked with * are mandatory.

          Save   Close   Edit My Shop

Figure 9–2. Create Shop page

4. Clicking the Save button creates your Shapeways Shop, as you can see in Figure
    9–3. The Shapeways Shop does not look that different from the "my designs"
    page, but there are few extra things you can do now when you open a shop.

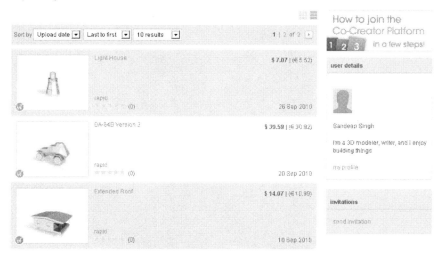

Figure 9–3. *Your Shapeways Shop*

5. Now that the shop is created, there are a few things you need to double-check
    to make sure they are correct.

6. Within each model's design detail page, change the shop properties of the
    model. Select the Available to All check box, and select "Show and allow
    ordering from the Model view state" drop-down list. This is to make sure that
    someone else accessing your store can purchase things. You can also select the
    categories, public galleries, and materials you want the model to be printable
    in.

7. On the bottom right of the page, enter a markup price for the model you are
    selling, as shown in Figure 9–4. Shapeways recommends you mark up the prices
    between 10 percent and 20 percent. Be careful, though, because raising the
    price too high might scare away your customers.

8. At the bottom, click Save Changes.

Figure 9–4. Making your profit

Using your custom URL link created earlier (Figure 9–2), you can access your shop from anywhere in the world. Now it's time to sell. Yes, that's right. The fun was all in the design process, but now the question is, will anyone buy your model? If you are looking to make some profit from the models you design, make sure to do some research before you design anything, and be ready to market your model. Tell your friends, family, and everybody about it. Post your designs on Facebook and on Twitter so others can see. From here it's all about being a salesperson. If people believe in your product and can see a use for it, then you are above the rest. I wish you all the best.

Google 3D Warehouse

The Google 3D Warehouse is a repository of SketchUp models. It's a great place to share your SketchUp files with others without having to convert it in any other format. The Google 3D Warehouse does not 3D print your models; you can only share your models and design models for Google Earth. 3D printing your models and selling them on Shapeways is your best option. Models in the 3D Warehouse are divided into two groups:

- Geo-referenced
- Non-geo-referenced

The difference between the two is that a geo-referenced models can be placed and referenced in Google Earth, while a non-geo-referenced model cannot be referenced on a map (we'll cover placing models in Google Earth later in the chapter). The great thing about the 3D Warehouse is that you can search and download these models. That's cool! To access the Google 3D Warehouse home page, visit http://sketchup.google.com/3dwarehouse/.

Downloading Models from the 3D Warehouse

To find models within the search bar, enter keywords describing the model you are looking for, or browse through the subcategories: 3D Building Collections, Featured Collections, Popular Models, and Recent Models. Once you have found a model, the next step is to view and download it. Click the image of the model that interests you. You will then be directed to the download page. Figure 9–5 shows the download page for a house model.

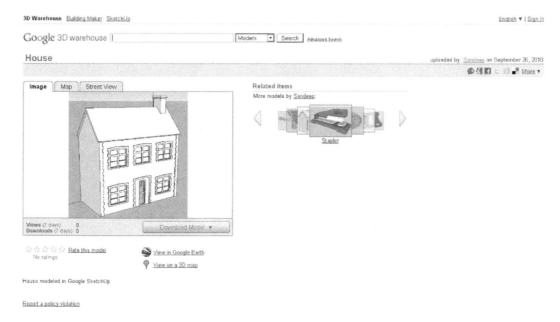

Figure 9–5. Google 3D Warehouse house model

Click the Download Model button (Figure 9–6). You then have the option of choosing two formats, Google Earth 4 (.kmz) and Collada (.zip). Download the Collada file to your computer, and open it with Google SketchUp to view the model. Some models can also be downloaded as Google SketchUp 7 (.skp) and Google SketchUp 6 (.skp) files.

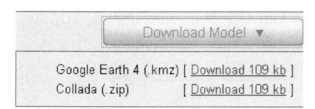

Figure 9–6. 3D Warehouse file types

Uploading Models to the 3D Warehouse

The great thing about the Google 3D Warehouse is the ability to share models. In this section, we will be going through the steps of uploading a design to the Google 3D Warehouse. You will first need a Google account.

1.   Visit the Google 3D Warehouse home page, and click the Sign In icon located in the upper-right corner of the web page. (You will find the Sign In icon on the upper-right corner of every Google 3D Warehouse web page.) You will then be directed to the login page (Figure 9–7).

Figure 9–7. Google login page

2.   Log in to Google Accounts with your user name and password. If you do not have a user name and password, click "Create an account now," and follow the instructions to create your own Google account.

3.   Once you are logged on, in the upper-right corner you will see a My Warehouse link. Clicking the link will display a drop-down menu with links to My Models, My Collections, and My Account. As a new user, you won't find any models under the My Models link. Clicking the My Collections link will direct you to the page shown in Figure 9–8. You will find information on the number of models you have uploaded, models that have been accepted for Google Earth, and the option of setting your preferences.

Figure 9–8. Google 3D Warehouse collections page

4.   Now, on the Google 3D Warehouse home page, click Upload. The Upload link is located in the upper-right corner of the page (Figure 9–9).

Figure 9–9. 3D Warehouse home page

5. This will direct you to the Upload to 3D Warehouse page (Figure 9–10). Make sure to fill in the form as specified. The file has to be in `.kmz` format, and the thumbnail has to be less than 2MB and less than 1200 by 1200 pixels. To export the SketchUp model as a `.kmz` file, select the File menu in SketchUp and then select Export ➤ 3D Model. Make sure to specify the file type as `.kmz`.

6. At the bottom of the page, select Publicly Viewable or Private. Also insert a thumbnail, title, and description of the model before uploading so people looking at your model on the 3D Warehouse will understand what they're looking at and are about to download. At the bottom of the page, click Upload. The model will then appear in your personal gallery.

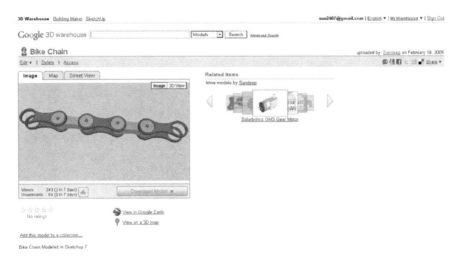

Figure 9–10. Model upload page in Google 3D Warehouse

Figure 9–11 shows a model of a bike chain uploaded into my personal gallery that I modeled using SketchUp. Now anyone in the world can search and download the model.

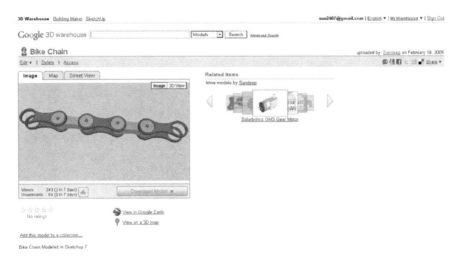

Figure 9–11. Bike chain model uploaded into Google 3D Warehouse

Google Earth

Google Earth is a great tool you can download and do all sorts of things with. The tool is great for searching places, finding driving directions, and designing models in SketchUp for Google Earth. All models are geo-referenced, so you can design your house and have it be considered for placement in Google Earth. It's just another great way of sharing some of your 3D models with the world. If you are planning to use SketchUp to develop models for 3D printing, then you probably are not going to use Google Earth. If learning how to draw models for Google Earth does not interest you, then by all means skip to the next section where we discuss the Thingiverse 3D modeling repository. If you're not sure what Google Earth is and would like to learn more about its features, then keep reading.

To get started, you will need to download a copy of Google Earth from http://earth.google.com. Currently version 5 of Google Earth is available for download. Locate the download link, and download a copy of Google Earth. Follow the on-screen instructions to install Google Earth. Once installed, Google Earth will open automatically, or you can simply double-click the Google Earth icon on your desktop to launch the program (Figure 9–12).

Google Earth

Figure 9–12. Google Earth desktop icon

When Google Earth opens, you will be presented with the screen shown in Figure 9–13.

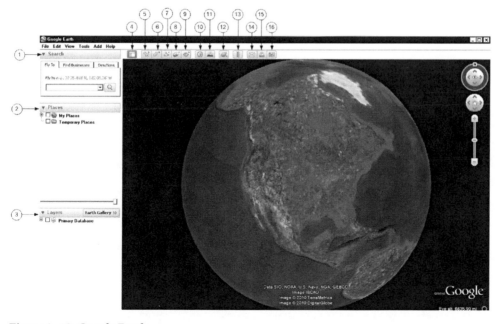

Figure 9–13. Google Earth

On the left of the Google Earth application are a set of panels. These and other tools are described in Table 9–1.

Table 9–1. Google Earth Interface

Name	Description
Search panel	Within the Search panel, you can type in a city, state, or country, and Google Earth will automatically find the location on the map.
Places panel	In this panel, you can save places you have visited in Google Earth and organize them.
Layers panel	In this panel, you can apply different features to your map: gallery, ocean, weather, traffic, and much more.
Hide sidebar	This hides the sidebar to get a full view of the window.
Add Placemark	You can add a placeholder within the map of interest for later reference.
Add Polygon	You can add polygons to the map.
Add Path	You can add paths on the map.
Add Image Overlay	You can add an image overlay on top of the earth.
Record a Tour	You can create a fly-by tour of places in Google Earth.
Show Historical Imagery	You can view images of places over time.
Show Sunlight Across the Landscape	Slide the cursor to observe the change in light for the entire day.
Switch Between Earth, Sky, and Other Planets	Switch between different maps in Google Earth.
Show Ruler	You can create a path or draw a line.
Email	You can e-mail an image in Google Earth.
Print	You can print the current map.
View in Google Maps	You can show the current view in Google Maps.

Making Movies

One of the cooler aspects of Google Earth is the ability to make movies of your favorite destinations. In this section, we will go through the steps of creating a movie by using the Record a Tour dialog box and by creating a path.

The moviemaking feature is a great way to present your models to others, especially if you are developing models for clients who need a visual of what the model might look like with the rest of the landscape. Once you have a video file recorded, you can play it without having to manually browse to each destination you'd like to display. You can use these file formats when making a movie:

- Windows Media Video (.wmv)

- Audio Video Interleave (.avi)

- Image Stream (.jpg)

- QuickTime (.mov)

For a detailed explanation of each file format, visit http://en.wikipedia.org/wiki/Container_format_%28digital%29.

Let's make a movie. The simplest way of making a movie in Google Earth is using the Record a Tour tool.

1. Click the Record a Tour icon, and the Record a Tour controls will appear (Figure 9–14).

Figure 9–14. Record a Tour controls

2. Before you even start recording, let's select a couple of places to tour in Google Earth (choose from Table 9–2). Type each destination into the Search panel, and press Enter. Observe how Google Earth zooms into each location.

3. Once you have all the locations entered into the Search panel, you are ready to make your movie. Click the red dot in the Record a Tour dialog box. Then approximately every five seconds, click the locations you entered into your Search panel.

4. Once you have zoomed through each location, click the Record/Stop button to stop the recording. After that, you will see another dialog box appear that will play your recording (Figure 9–15).

Figure 9–15. Playing the tour

5.   The dialog box allows you to play, fast-forward, rewind, refresh, and save your tour. After the tour has finished playing and you are happy with the results, then go ahead and save the tour. Give your tour a name, and then click OK. To play your tour again, click the camcorder icon within the Places panel (Figure 9–16).

Figure 9–16. Places selection

Table 9–2. Destinations Around the World

Name	Description
Rio de Janeiro	City in Brazil known for its beautiful beaches and the famous statue of Jesus Christ overlooking the city.
Eiffel Tower	Located in Paris, France. Considered to be one of the seven wonders of the world.
Amritsar	A city located in the northern part of India in the state of Punjab. Amritsar is well known because of the Harmindar Sahib, also called the Golden Temple.
Taj Mahal	Located in Agra, India. Considered to be one of the seven wonders of the world.

Another way to create a tour in Google Earth is by creating a path. Figure 9–17 shows the Golden Gate Bridge in San Francisco.

Figure 9–17. *Arial view of the Golden Gate Bridge*

To create a path from one end of the bridge to the other, follow these steps:

6. Click the Path button. The New Path dialog box will appear.

7. Give the path a name. You can also describe the path and choose the path color, view, and altitude options (Figure 9–18).

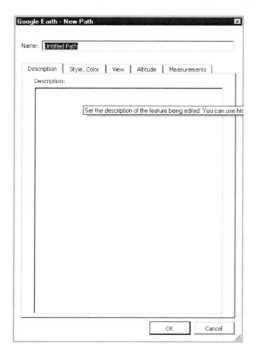

Figure 9–18. *New Path dialog box*

8. Click one end of the bridge, and trace a line to the other. Once you are done, click OK within the New Path dialog box to accept the addition.

9. Now that you have created a path, you can play it. Select the path created. At the bottom of the Places panel is the Play Tour button. Click the Play Tour button to see your animation play. If you do not want to scroll throughout the image, you can always double-click the location you want to zoom to. Every time you double-click, you will zoom closer into that location.

About Layers

In Table 9–1, I briefly mentioned the Layers panel. With the Layers panel, you can add a tremendous amount of information to Google Earth. Figure 9–19 shows a list of current layers you can choose in Google Earth.

Figure 9–19. Layers panel

Table 9–3 provides a brief description of the first layers. Select any of them, and select among a group of subheadings.

Table 9–3. Layers Within Google Earth

Name	Description
Borders and Labels	See the outline of countries, states, counties, and their labels.
Places	Create a layer for planning a trip that lets you take a look at the available amenities in the area: bars, dinning, lodging, gas, and pharmacies are some of the many options.
Panoramino Photos	See photos from the world tagged to their geographic location.
Roads	View freeways, highways, and residential roads.

Name	Description
3D Buildings	View models in 3D.
Ocean	View what's underneath the ocean.
Street View	View Google Earth at street level.
Weather	Track weather conditions with this real-time weather tracking system.
Gallery	Get all kinds of information about the world.
Global Awareness	Learn about different global awareness programs.
More	Find resources to more information you can find in Google Earth.

Measuring Distances

If you're a real estate agent, land buyer, or builder, then the Ruler tool might be for you. It's great for mapping distances of land or planning your next hike.

1. Click the Show Ruler button in Google Earth. The Ruler dialog box will appear (Figure 9–20).

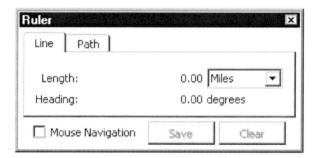

Figure 9–20. Ruler dialog box

2. With the Line tab selected, you can draw straight lines in Google Earth. Select the Path tab, and you can trace a path that has twists and turns. In Google Earth, zoom into a location on the map, and select the Path tab.

3. Trace a path within the map. Every time you click your mouse, a new line is drawn attached to the previous line. A red dot represents the starting point, and as you click, more dots are created defining your path. To erase the line, right-click your mouse. In Figure 9–21, I have traced a simple path using the Ruler tool.

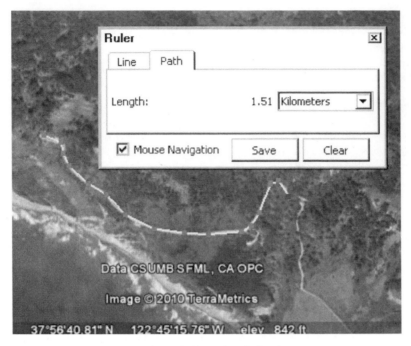

Figure 9–21. Tracing a path with the Ruler dialog box

Notice that the Ruler dialog box gives me the distance of my path. It measure 1.51 kilometers.

Placing Models in Google Earth

One of the interesting things that makes Google Earth such a great tool is that you are not limited to only viewing maps and models; you can also upload your own models. To place models in Google Earth, you have to go through a couple of steps. The first thing to decide is what structure you want to model in Google Earth. It could be your house, school, university, medical center, bridges, or dams. Make sure your model is geo-located, because the structure has to be fixed on the earth. Homes, bridges, schools, and medical centers are all in a fixed in one location, whereas cars, trucks, people, and boats are not. Google SketchUp does not stop you from modeling non-geo-located objects, and you can still share them through the Google 3D Warehouse, just not with Google Earth.

1. Open Google SketchUp, and click the Add Location icon. The Add Location dialog box will appear, as shown in Figure 9–22. Find an object on the map you want to model.

Figure 9–22. Top-down view of the two buildings that will be constructed in SketchUp

2. Zoom into the location of the object you plan to model, and click Select Region. A square will appear with four blue pins on each corner, as shown in Figure 9–23. Select each corner, and try to get a close-up of the image. Then select Grab in the upper right of the dialog box.

Figure 9–23. Cropping the image with the four blue pins

3.   This will import the current view of the image into SketchUp (Figure 9–24). Using this image, construct the model.

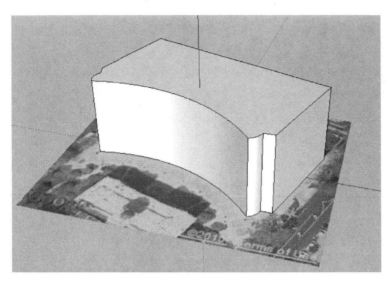

Figure 9–24. Current view in Google SketchUp

After you are done building your model, the next step is to import your Google SketchUp model into Google.

Now would be a good time, if you haven't already done so, to zoom around the model to make sure it is aligned perfectly with the photograph. If you are planning to upload the model to Google Earth, you should keep the model as simple as possible. Larger file sizes will slow down the operation of Google Earth. Larger files mean a large number of faces that have to be uploaded in Google Earth. Figure 9–25 shows a square, rectangle, circle, and star. All of these shapes have a single face.

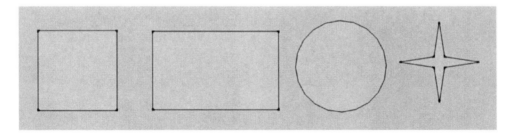

Figure 9–25. Square, rectangle, star, and circle, all single-faced

But guess how many faces would be created if we extruded each one of these shapes? If you guessed 44 total faces, then you are exactly correct. The square will have 6 faces, the rectangle will have 6 faces, the circle will have 22 faces, and the star will have 10 faces, all of which is illustrated in Figure 9–26. Notice that with just an extrusion there is a significant increase in the number of faces in each model.

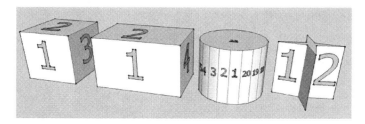

Figure 9–26. Number of faces on extruded models

To keep your models simple, use photographic projections instead of modeling each and every detail of the model. The image will be much more realistic, and it will cut your modeling time in half. When modeling objects, draw only the outer edge of the building, and leave the internal details behind. Adding details such as windows into the model will increase the number of faces and edges, increasing the file size of the model.

Now, to upload your model into Google Earth, click the Preview Model in Google Earth icon (Figure 9–27).

Figure 9–27. Place Model icon

After importing, Figure 9–28 shows what the model looks like in Google Earth. Only you can see the model in Google Earth. To share your model with others, you will have to upload your design into the Google 3D Warehouse. Then Google decides whether your model is suitable for placement in Google Earth.

Figure 9–28. Placing model within Google Earth

What if you wanted to see someone else's model in Google Earth?

4. In the left panel of Google Earth under Layers, select 3D Buildings. If you are looking at the map from the top down, the 3D buildings will not be visible. You will need to orient the map to its side before the buildings will appear.

5. From the View menu, select Show Navigation and then Always. The navigation tool will appear as shown in Figure 9–29b. Using the tilt navigation controls, tilt the surface of the map so it is at eye level. Notice as the map tilts, the buildings start to appear (Figure 9–29a).

6. Select the 3D Buildings layer once more to see all the buildings disappear.

a.

b.

Figure 9–29. (a.) View 3D Buildings in Google Earth; (b.) navigation tool

Wasn't that awesome? You just viewed 3D Warehouse models in Google Earth. Now it's your turn. Draw your house, upload it into the 3D Google Warehouse, and share it with the world.

Thingverse: Digital Design for Physical Objects

Thingiverse (Figure 9–30) is a recent development in the world of 3D modeling. Thingiverse offers the ability for users to upload and share models and is similar in many ways to the 3D Warehouse but is a little different. Unlike the 3D Warehouse, you can upload and download models that can be physically built with your own laser cutter, 3D printer, or CNC machine. The site caters to the hobbyist who enjoys building things. For more details about the site, visit www.thingiverse.com.

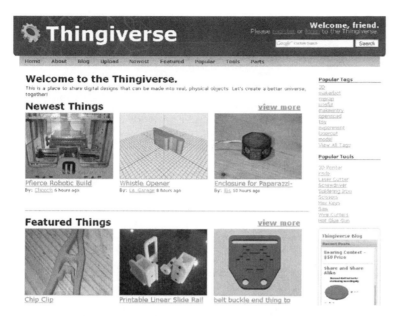

Figure 9–30. Thingiverse web site

Thingiverse has nine links that make up the page: Home, About, Blog, Upload, Newest, Featured, Popular, Tools, and Parts.

- Clicking the Home link will direct you to the home page of Thingverse.

- About gives you some detail on Zach Hoeken and Bre Pettis, the founders of Thingiverse.

- Click Blog to get upcoming news in the world of 3D printing.

- Clicking Upload will direct you to the upload page where you can upload your own model.

- Clicking Newest will display all the new models uploaded to Thingiverse.

- Click Featured to see a collection of featured models.

- Click Popular to see some of the interesting models people have uploaded

- If you're interested in learning about some of the tools people are using to design their models, click the Tools link.

- Under Parts, create a listener of all the parts that make up your model for easy tracking.

Registering for an Account

Before you can download or upload a model, you first need to register for an account with Thingiverse. Not to worry at all—signing up is free. On the upper right of the Thingiverse web site, click "register." The Register a new Account page will appear, as shown in Figure 9–31.

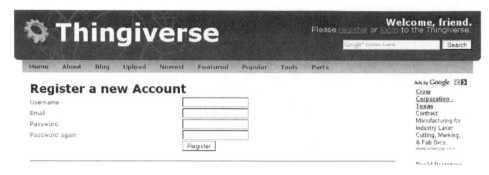

Figure 9–31. Thingiverse account registration

To register, enter a user name, e-mail, and password, and then click the Register button. The Edit Profile page will appear with a selection of tabs: Details, Profile Image, My Tools, Email, and Back to Profile, as shown in Figure 9–32.

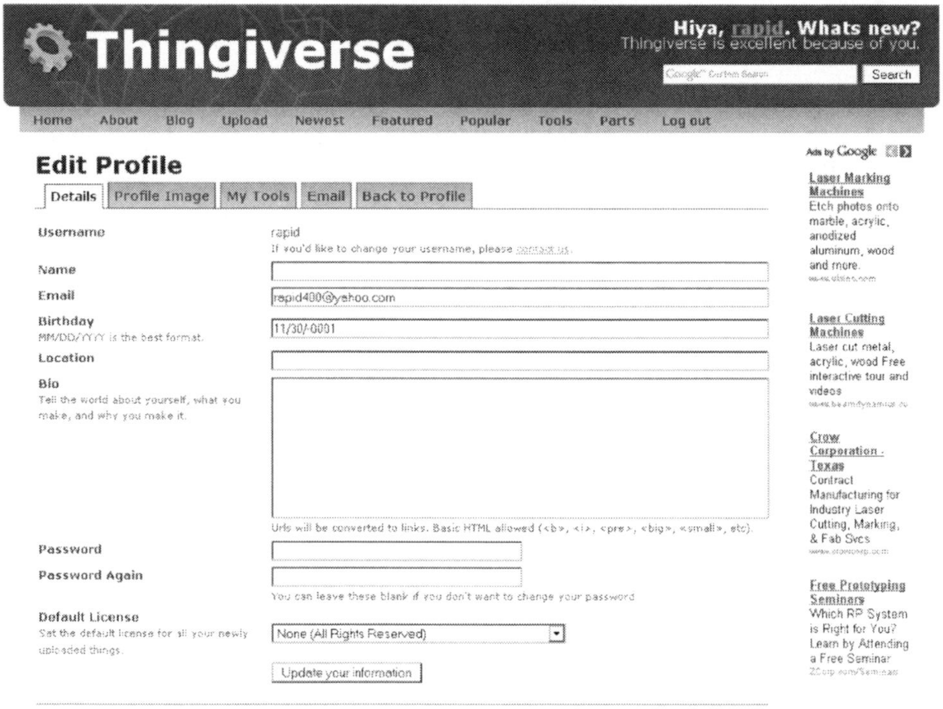

Figure 9–32. Edit profile page

- On the Details tab, you can change or add your name, e-mail, birthday, location, bio, and password. At the bottom of the page is an option to select a license for your model. This will be the default license for every model you upload. You can always select a license after uploading your model.

- On the Profile Image tab, you can add an image of yourself.

- The My Tools tab allows you to select an assortment of tool you possess. Simply select the tool you have, and it will be automatically added into your repository.

- On the Email tab, select what types of messages you want to receive.

- Your About page will display all the information you have added into these pages.

Uploading a Model to Thingiverse

To upload a model into Thingiverse, click the Upload link from the main navigation bar. The Create New Thing page will appear, as shown in Figure 9–33.

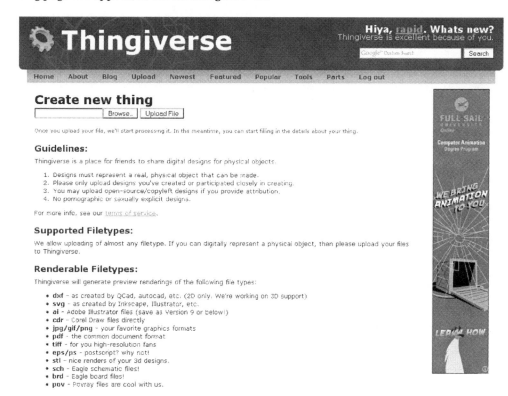

Figure 9–33. Thingiverse upload page

To upload a file, there are a few guidelines to follow and a selection of file types that Thingiverse will allow you to upload. Make sure the designs you make are real objects that can be manufactured, make sure to upload only the designs you created, and avoid uploading any pornographic material. There are only a select set of file types that Thingiverse will allow for upload:

- `.dxf`
- `.svg`
- `.ai`
- `.cdr`
- `.jpg/.gif/.png`
- `.pdf`
- `.eps/.ps`
- `.stl`
- `.sch`
- `.brd`
- `.pov`

Unfortunately, Thingiverse does not accept `.skp` or `.dae` files for upload.

No worries, because this is where you can use the handy-dandy CADspan software to convert your files, as you saw in Chapter 5. Once you have the file in an acceptable format, follow these steps:

1. Click Browse, and upload the file. As an example here, I'm using the model of the house we constructed in Chapter 7 and have converted it to an `.stl` file for upload.

2. After adding the model, click the Upload File button. The Edit Your Thing page will appear with a selection of seven tabs, as shown in Figure 9–34.

Figure 9–34. Edit upload page

3. Select the Detail tab, and enter the following information:

    - For the name, enter **House**.

    - For the description, enter **Model of a house created in Google SketchUp using Match Photo. The file was converted using CADspan to an stl file.**

    - You can enter instructions for the design of your model or leave it blank.

    - Select the license type as Attribution – Creative Commons. For assistance in deciding what types of license is best for your model, visit CreativeCommons.org. The Creative Commons license allows you to add specific restrictions on your model based on the rules of copyright.

4. On the Images tab, browse and select an image of the house to import. Click Open and then Upload.

5. Select the Tag tab, and select or type keywords describing the house model.

6. On the Tools tab, select the tools you used to construct your model.

7. On the Parts tab, select all the related parts that make up your model.

8. After you are all done, click Back to Thing. Now your model is ready to be shared with the rest of the world, as shown in Figure 9–35.

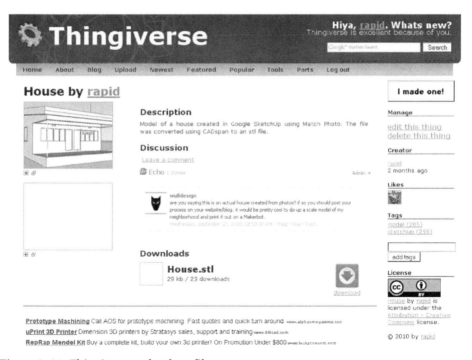

Figure 9–35. Thingiverse upload profile page

225

If at a later date you want to edit or delete the model, select "edit this thing" or "delete this thing" on the right side of the menu bar. That's it—now your model is on Thingiverse for others to download and 3D print. Remember to publish your model so that others can see and download it in Thingiverse.

Summary

This chapter was all about showing you how you can share and sell your model with the rest of the world. We covered opening a Shapeways Shop, uploading models to the 3D Warehouse and Thingiverse, and modeling for Google Earth. In the next chapter, you'll learn about SketchyPhysics and how it can be used to animate your models.

■ ■ ■

CHAPTER 10

Animate with SketchyPhysics

SketchyPhysics is an interesting plug-in developed by C. Phillips that's used for animating models in SketchUp. In this chapter, you'll use SketchyPhysics and see several different examples demonstrating the features it has to offer. You'll start by getting an introduction to the SketchyPhysics tool set. Then you'll learn how you can animate models with the tool set. You'll learn how your models interact with other models, you'll learn how to implement game controller functionality in your models, and you'll learn more about the SketchyPhysics UI module. You'll conclude the chapter by animating the armored car model designed in Chapter 8.

Getting to Know SketchyPhysics

Why in the world would you want to animate your models if all you need to do is 3D print them? For starters, animations allow you to easily show the purpose of the model. But the biggest benefit of animating a model is to test the model's form, function, and performance before even sending it for 3D printing. For example, if you were designing a bicycle, you would want to see how the wheels turn, how the handle bar rotates, and how the chain moves. From this, you can easily deduce any flaws in the form, function, and performance of the model.

SketchyPhysics allows you to debug a model for any errors, which, if detected early enough in the design process, will save you time and money. In this section, you'll see how to download and install the plug-in and then get a crash course in some of its more useful features.

Installing SketchyPhysics

To download your own copy of the plug-in, visit the SketchyPhysics web site at http://code.google.com/p/sketchyphysics/. Select the Download tab, and then select SketchyPhysics 3.1 Windows Installer (Figure 10–1).

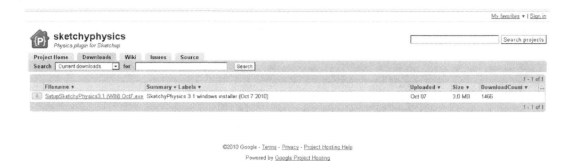

Figure 10–1. SketchyPhysics download page

After downloading the installation file, double-click it to install the plug-in (all the files will be automatically placed into the plug-in's folder when you install it).

Setting Up the Tool Sets

When you open SketchUp, you'll notice that there are now four additional tool sets in your toolbar: SketchyPhysics, SketchyPhysics Joints, Sketchy Solids, and Sketchy Replay. If none of these tool sets is present, then you will need to activate them manually (Figure 10–2).

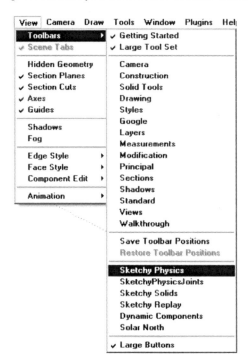

Figure 10–2. Activating SketchyPhysics tool set

Select View ➤ Toolbars. Then select each toolbar you want to open individually: SketchyPhysics, SketchyPhysics Joints, Sketchy Solids, and Sketchy Replay. Now let's review these tool sets as a primer to the "Learning by Example" section.

SketchyPhysics

Figure 10–3 shows the SketchyPhysics toolbar, while Table 10–1 describes what these tools do.

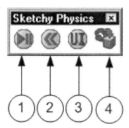

Figure 10–3. SketchyPhysics toolbar

Table 10–1. Four Buttons on the SketchyPhysics Toolbar

Name	Description
1. Play/Pause Physics Simulation	Runs your SketchyPhysics simulation
2. Reset Physics Simulation	Restarts SketchyPhysics simulation
3. Show UI	Displays the user interface
4. Joint Connector	Connects parts of a model to create a joint

Sketchy Solids

The Sketchy Solids tools shown in Figure 10–4 allows you to model 3D shapes and predefined joints, as described in Table 10–2.

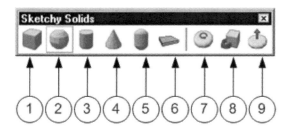

Figure 10–4. Sketchy Solids toolbar

Table 10–2. *Tools in the Sketchy Solids Toolbar*

Name	Description
1. Box	Models a box
2. Sphere	Models a sphere
3 .Cylinder	Models a cylinder
4. Cone	Models a cone
5. Capsule or chamfer	Models a capsule
6. Solid floor	Automatically generates the surface of the model
7. Create Wheel	Draws a wheel that will rotate
8. Create Door	Creates a block that can operate similar to a door
9. Create Lift	Creates a block that can slide

SketchyPhysics Joints

The SketchyPhysics Joints tool set consists of 11 tools (Figure 10–5). A description of each, starting from the left, is provided in Table 10–3.

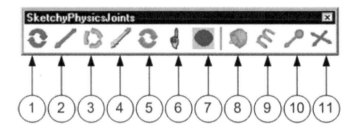

Figure 10–5. *SketchyPhysics Joints toolbar*

Table 10–3. Ten SketchyPhysics Joints

Name	Description
1. Hinge Joint	Objects will spin around the joint's center when linked with the joint.
2. Slider Joint	Objects will move up and down linearly.
3. Servo Joint	This is a joint that can be controlled in different directions.
4. Piston Joint	This is like the slider joint, but it can be controlled.
5. Motor Joint	This is also a controlled joint designed for wheels.
6. Gyro Joint	Objects connected with this joint will operate as a gyro.
7. Fixed	Fix objects to other objects.
8. Corkscrew Joint	Objects using this joint will spin around the joint's center.
9. Spring Joint	This acts link a spring, bringing the joint back to its original position.
10. Ball Joint	Objects will move around the joint.
11. Universal Joint	Objects can rotate in any direction except their axes.

Sketchy Replay

Sketchy Replay is a new toolbar and part of SketchyPhysics 3 (Figure 10–6). With this toolbar, you can record a series of movements in SketchyPhysics and play them back, as described in Table 10–4.

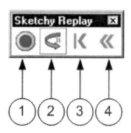

Figure 10–6. SketchyPhysics Replay toolbar

Table 10–4. *Sketchy Replay*

Name	Description
1. Toggle Recording	Starts and stops recording of animation
2. Play Animation	Plays the recorded animation
3. Rewind to First Frame	Rewinds the animation
4. Reverse	Animation plays in reverse

Learning by Example

To better familiarize yourself with the different tool sets, you will see a few examples demonstrating the use of each one. You will first be introduced to the SketchyPhysics UI module. Then you'll go through a simple example to learn how to animate a fan blade and post. In the "Animating a Sphere" section, you will learn about the different properties that objects in SketchUp can have when using SketchyPhysics. Then for fun, you will construct a maze game by applying game controller functionality. You'll also learn about applying gravity to models with the design of a hockey table. In the last two sections of the chapter, you'll construct a simple shooting game and animate the armored car you designed in Chapter 8.

A lot of examples are covered in this section, but don't be alarmed. The goal simply is to introduce to you another tool that you can use to test the functionality of the models you design in SketchUp and for 3D printing. Read these sections to get familiar with the some of the basics of SketchyPhysics. Along the way, I will refer you to couple of sites you can use to learn more about the tool.

SketchyPhysics UI Module

When an object or joint is selected in the model, the UI module will display all of its parameters. The UI window is divided into four categories: Joints, State, Properties, and Shapes. The dialog box in Figure 10–7a is empty. When you select a joint or object in SketchUp, then its properties will be displayed in the dialog box, as shown in Figure 10–7b.

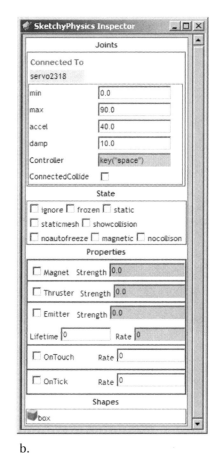

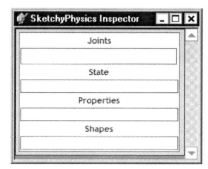

a.   b.

Figure 10–7. SketchyPhysics Inspector dialog box

When an object is selected under Joints, you will see all the joints attached to the object. Under State, you can select the state of the model when you run your animation. Under Properties, you have the option of changing a few of SketchyPhysics advanced functions; these are Magnet, Thruster, and Emitter. SketchyPhysics Inspector allows you to add scripts to produce different effects in your model. These commands are simple snippets of code that you can add into the boxes of the controller (present only when you select one of the joint connectors): Magnet, Thruster, and Emitter Strengths. To learn more about scripting for SketchyPhysics, visit http://sketchyphysics.wikia.com, and select Ruby.

Animating a Fan Blade and Post

In this example, you'll learn how to use the Hinge joint and Joint connector to animate a simple model of a fan blade and post. This is a three-step process. First you need to create a floor. The floor will be the surface on which the model is placed. Without the floor, the model would be in a free-fall when you simulated the animation. Then you'll model the post and fan blade. After all the parts are constructed, you will run the animation and test how it all works. So, let's get started!

Constructing the Floor, Post, and Fan Blade

To construct the floor, post, and fan blade, follow these steps:

1. Click the "Create a solid floor" button in the Sketchy Solids toolbar. You can also draw the floor using the Line tool: draw a surface, and use the Push/Pull tool to create a box. Group the box, right-click, and from the drop-down menu select SketchyPhysics ➤ Shape ➤ staticmesh. You have just created the floor of the model.

2. Model a post and fan blade, as shown in Figure 10–8a, using the Line, Rectangle, and Circle tools. Make sure to draw the model on top of the floor you just created. And make sure it isn't bigger than the floor.

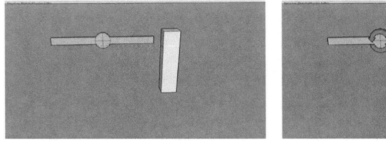 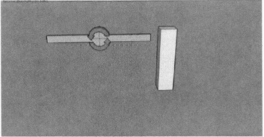

a.   b.

Figure 10–8. (a.) Fan blade and post; (b.) Fan blade with Hinge joint

3. After drawing the fan blade and post, group them separately. All objects must be grouped before they can be animated using SketchyPhysics.

4. From the SketchyPhysics Joints toolbar, select the Hinge tool, indicated by the circular arrows. Click the center of the fan blade, and click a second time, making sure to orient the center of the hinge so that it is perpendicular to the blade (Figure 10–8b). If the hinge appears small compared to the entire model, you can enlarge it using the Scale tool.

5. From the SketchyPhysics toolbar, click the Play/Pause button. The animation will run, and the model will fall apart. This is actually normal. If the fan blade does not fall over, then the state of the blade must be set to static. In this case, right-click the blade from the drop-down menu, select SketchyPhysics ➤ State, and deselect static. Refer to Table 10–6 later in this chapter for all the states of an object in SketchyPhysics.

6. Now click the Reset icon, and all the objects will return to their original places. Before you continue, group the fan blade with the hinge together. (In step 3, you grouped only the individual parts.) Then place the fan blade perpendicular to the post (Figure 10–9a).

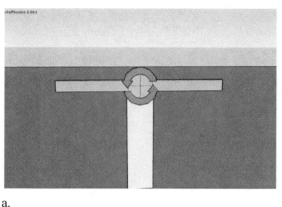 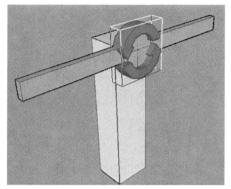

a.   b.

Figure 10–9. Joint connecting the hinge to the post

The next step is tricky; the blade and hinge will be joint-connected to the post.

7. Select the Joint Connector tool from the SketchyPhysics toolbar, and then select the post, hold Ctrl on your keyboard, and then select the hinge. You'll know that everything is hinge connected when you see a yellow box around the hinge and a green box around the post (Figure 10–9b).

8. Hit the Play button, and use your mouse to apply some momentum to the blade. The blade will move in the direction you provide momentum. You will notice that the post tips over with an increase in momentum.

9. Right-click the post, and select SketchyPhysics ➤ State ➤ Static. This will lock the post in its location.

10. Now that the model is step up correctly, the next step is to add a slider to control the fan blade in the model.

Testing the Animation with the Slider

Before jumping in and adding the slider, run the animation once more to see how it operates. You should have a spinning blade. Isn't that awesome? You just animated your first model using SketchyPhysics. Now let's take a look at how you can control the fan blade using the UI module and the slider I was talking about.

1. To open the UI module, select the Show UI from the SketchyPhysics toolbar. Select the Hinge joint in the model. The UI module then presents the properties of the hinge (Figure 10–10).

Figure 10–10. *SketchyPhysics Inspector dialog box*

The SketchyPhysics Inspector dialog box, or UI module, is filled with an assortment of options. Under Joints are the following menu options: min, max, accel, damp, and controller. By changing these menu options, you can change how the joint operates. The "min" and "max" settings represent the angle of rotation in the model.

2. Type in **-90** and **90** for your min and max values, and within the Controller box, type **slider("blade")**. Then hit the Simulation button.

The SketchyPhysics-Slider dialog box will appear, showing a slider and its name. Moving the slider left or right will cause the blade to move +/- 90 degrees (Figure 10–11). The slider shows a number between 0 and 1 when dragging, which indicates the two extremes.

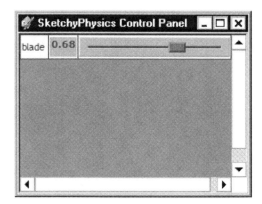

Figure 10–11. *SketchyPhysics-Slider dialog box*

By adding more joints to the model, you can increase the complexity and movement capability of the model. You can also add sliders to the model. There are great examples online of how others have utilized SketchyPhysics functionality. Table 10–5 lists some links to videos on YouTube, which you might find helpful.

Table 10–5. Links to Videos Demonstrating Use of the SketchyPhysics Plug-in

Name	Description
Robot arm, SketchUp plan #1	`www.youtube.com/watch?v=d2OBCpbCEc4`
Segway (SketchyPhysics 2)	`www.youtube.com/watch?v=kvOvO5cW6K4`
Walking Robot (SketchyPhysics 2)	`www.youtube.com/watch?v=OOOGKPxS-So`
SketchUp Examples	`www.youtube.com/watch?v=qLvXvzPvfWI`
Star Wars Vs. SketchUp	`www.youtube.com/watch?v=sN8BYT59h5A`

Now I don't expect you to design advanced SketchyPhysics animations from day one. The process will take some time, but I'm sure you can design some of the more advanced animations with practice.

Animating a Sphere

Setting the State and Shape of a model is very important when animating with SketchyPhysics. Setting the Shape of a model affects the way it interacts with other objects. Setting the State lets other objects know what it is they're interacting with. In this section, we will be modeling a sphere and defining its Shape with a mesh. You can think of a mesh as a net, and this net describes the shape of the model such as the skeleton forming the shape of the human body.

For example, here's how:

1. Using the Sketchy Solids toolbox, draw a sphere (Figure 10–12).

2. Using Sketchy Solids automatically groups the sphere. If you were drawing it from scratch, you would need to group the sphere before SketchyPhysics would allow you to animate the model.

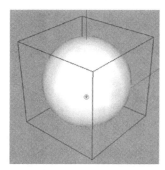

Figure 10–12. SketchyPhysics sphere, grouped automatically

3. Within the SketchyPhysics toolbar, click the Play/Pause physics simulation button to see the ball fall down.

With your mouse, drag and pull the sphere, and watch it move. Now exit the animation. To change the properties that define the sphere, right-click and select SketchyPhysics from the drop-down menu (Figure 10–13).

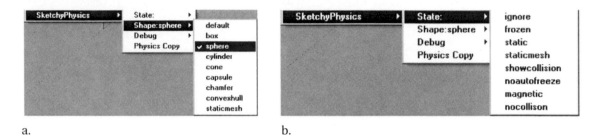

a.　　　　　　　　　　　　　　　　　　b.

Figure 10–13. State and shapes properties

Within the drop-down menu there are several options. These are State, Shape, Debug, and Physics Copy. Table 10–6 describes the function of each of the states you can apply to a model.

Table 10–6. Different States Within SketchyPhysics

Name	Description
Ignore	The model will act as a ghost, remaining invisible to other objects.
Frozen	The model will move only once another object has touched it.
Static	The model will stay in place, and other objects will bounce off it.
Showcollision	This shows the mesh that the model is defined by.
Magnetic	This will stick to anything magnetic.

4. Using the Shape option in SketchyPhysics allows you to specify the shape other objects will encounter when coming into contact with your model. Make sure the model of the sphere you constructed in step 1 is defined by a sphere.

5. Right-click the model, and select SketchyPhysics ➤ Shape ➤ sphere.

If you had manually constructed the sphere, it would have been defined as a box. Table 10–7 lists all the other available options within the Shapes list.

Table 10–7. Different Mesh States Within SketchyPhysics

Shape	Description
Default	Assigns a default mesh to the object
Box	Assigns a box mesh to the object
Sphere	Assigns a sphere mesh to the object
Cylinder	Assigns a cylinder mesh to the object
Cone	Assigns a cone mesh to the object
Capsule	Assigns a capsule mesh to the object
Chamfer	Assigns a chamfer mesh to the object
Convexhull	Assigns a convexhull mesh to the object
Staticmesh	Applies a static mesh, keeping the object in place

6. Select Debug, and make sure Readback Collision Geometry has been selected (Figure 10–14a). Now you are ready to run the simulation.

7. From the SketchyPhysics toolbar, click the Play/Pause Physics simulation button. You should see the sphere drop down, leaving behind a wire mesh representing the sphere (Figure 10–14b). If this is the case, then you have set up the model correctly.

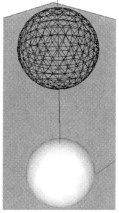

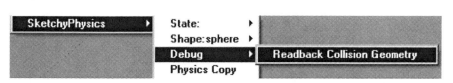

a.   b.

Figure 10–14. Mesh defining the structure of the sphere

With the sphere mesh, other models will interact with it as a sphere. Sometimes the model can be defined by a different mesh, which could result in behavior of your model that you did not intend. The following example shows what a wrong mesh might look like.

Wrong Mesh

There will be times when you're working with a model that has not been assigned the correct mesh. Let's look at another example and see how SketchyPhysics defines the model. In Figure 10–15, I have drawn a pyramid.

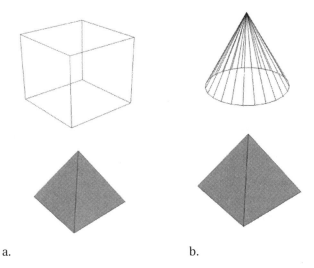

a.   b.

Figure 10–15. (a) Pyramid defined by cube mesh; (b) pyramid defined by cone mesh

After running the simulation, notice that the pyramid is defined by a box mesh (Figure 10–15a). Other objects will interact with the pyramid structure as a box. To change the mesh, right-click the object, and select Shape:cone ➤ cone. Now run the animation over again. What you should see is the pyramid defined by a cone mesh (Figure 10–15b). The cone is not an exact fit for the model, but it is close enough to do the job.

Adding Game Controller Functionality: the Joystick

I'm not expecting you to make games in SketchyPhysics, but this is such a cool feature that I had to introduce it. So far in this chapter you have been controlling your models with the mouse. The great thing about SketchyPhysics is that you can also add joystick/game controller functionality. With a joystick (Figure 10–16), you can control the Hinge, Slider, Piston, Servo, and Motor joints. Later you will also learn how you can use the joystick to create a game in SketchyPhysics.

Figure 10–16. *USB game controller*

With the joystick/game controller, you can control the function of the animation in the left, right, up, and down directions. Let's create a simple animation to test the functionally of the game controller. If you do not have a game controller, you can buy one from your local electronics store. If you're looking for a cheap controller, visit your local secondhand store. I'm sure you will be able to find one for a few dollars. The game controllers are very simple to install. They start functioning immediately when plugged into the computer. Just to double-check and make sure that your game controller is working, look at the game controller functions within the Windows Control Panel.

From the Start menu, select Control Panel. From the Control Panel, double-click and open the Game Controllers icon. With this dialog box, you will be able to control and calibrate your game controller (Figure 10–17).

Figure 10–17. *Game controller settings*

If your game controller is not present, then a driver might need to be installed for operation. Check the box or the manufacturer's home page for details on your particular controller. Once everything is in working order, the next step is to create an animation using the functionality of SketchyPhysics.

1. In SketchUp, construct the model shown in Figure 10–18. The surface on which the ball will role is the solid floor. The bars that separate each section of the maze were drawn using the Create a Box tool in SketchyPhysics.

2. Right-click each bar, and from the drop-down menu select SketchyPhysics; then choose "static" as the state. This will lock each of the bars in place.

3. Group the floor and boxes together.

4. Now place hinges on the left and back side of the floor.

tchyPhysics3 RC1

Figure 10–18. Maze game

5. Type **joy("leftx")** into the Controller box for the hinge connector on the left of Figure 10–19a. Type **joy("lefty")** into the Controller box for hinge connector at the back of Figure 10–19b. Remember that to access the parameters of each Hinge you will have to select them.

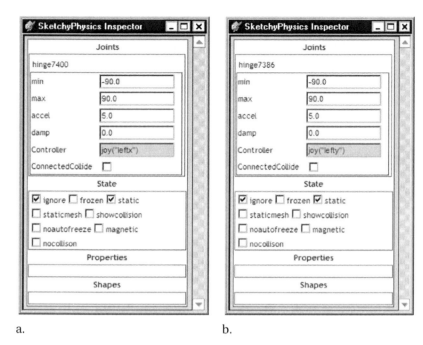

a.   b.

Figure 10–19. SketchyPhysics Inspector boxes

6. For both hinge connectors, enter **-90** for min, **90** for max, and **5** for accel.

7. To view the controller box, select the Hinge, and select Show UI. If you don't have a joystick, then try making it with a slider. Enter the following statements into the Controller box to create a slider: **slider("leftx")** and **slider("lefty")**. Also, remember to create a sphere and box in which the ball can fall.

8. Run your animation using the game controller move the floor, and guide the ball into the box.

Wasn't that exciting? You just created your first game in SketchyPhysics. If you got stuck along the way, try going through the steps a second time. Learning to animate can be very difficult, especially if this is your first time. If you still have trouble, refer to the SketchyPhysics wiki page at `http://sketchyphysics.wikia.com/wiki/SketchyPhysicsWiki`.

Using a Hockey Table and Puck to Simulate Gravity

In this section, you will be designing a hockey table and using the gravity options in SketchyPhysics to control the behavior of your puck.

1. First construct a model of air-hockey table, as shown in Figure 10–20.

2. Then model a puck. All you need is to model a thin cylinder.

3. Remember to group the table and puck separately, and also set the properties of the table as static. After creating the table and the puck and then running the simulation, notice that the contact between the puck and table results in friction. As a result, the puck does not move, or barely moves, in any direction. How can you reduce the friction between the table and puck? There are several methods to solve this problem. Changing the design of the puck would make a difference. Making a ball bearing design could surely help in this situation.

Figure 10–20. Air-hockey game

But, there is actually an easier way to get around all this, and that is to reduce gravity in our model.

4. In the Plugins menu, select SketchyPhysics and then Physics Settings. The Physics Setting dialog box appears (Figure 10–21).

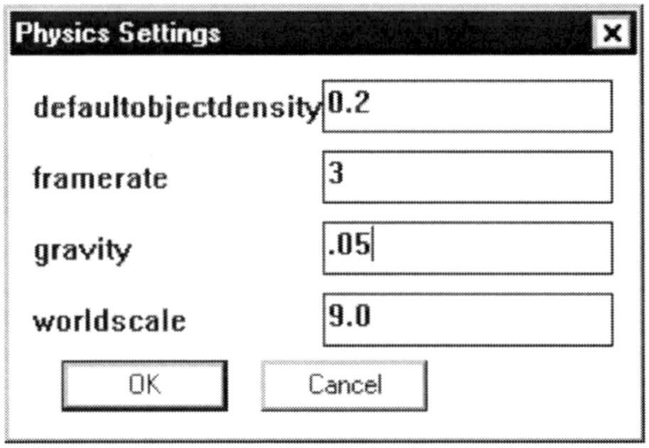

Physics Settings   ✕

defaultobjectdensity `0.2`

framerate `3`

gravity `.05`

worldscale `9.0`

[OK]   [Cancel]

Figure 10–21. Physics Settings dialog box

Increasing the gravity value increases the gravity in SketchUp. A value of 1 is set as the default value within SketchUp.

5. To simulate weightlessness in the air-hockey table, you simply reduce the gravity to .05, click OK, and run the simulation. If the puck flies through the side of the model, check the puck's State (refer to Table 10–6).

Notice that the hockey puck is now moving at a faster pace than before. Reducing the gravity even further will cause the hockey puck to fly off the table. You will need to find the right equilibrium to meet your needs. There are also other useful options located within the Physics Settings dialog box.

Framerate is the rate at which each frame changes. Smaller values will create faster changes in the number of frames and result in a smoother simulation. Higher framerate settings will result in a jumpy simulation. Worldscale is an interesting option. You can think of it as a scale parameter. If the object does not meet or exceed the designated parameters, the object will stay static or move very slowly compared to the worldscale. Think of the earth as an example for a moment. The earth rotates around its axis at approximately 1670km/hr. Compared to this rate, humans walking on the earth would seem to be static. From a large worldscale, the movement of smaller object becomes inconsequential, while from a small worldscale, even the tiniest of objects can have significant movement. To test the worldscale function, draw a small object and a large object, and experiment with altering your worldscale settings.

Modeling a Shooter for Animation

In this section, you will be modeling a shooter. There are a couple of ways you can design the shooter. You can use the Slider, Piston, or Emitter to design the shooter. To develop an automatic shooter in SketchUp, you'll be using the Emitter, because it is probably the easiest way to design a shooter in SketchUp.

Before you dive ahead and create the shooter, let's first understand how the Emitter options work within SketchyPhysics. The Emitter box defines the strength at which the object fires. Setting this value to zero will create copies in place. Lifetime is the length of time that each copy will last based on the number of frames. Rate is the number of frames within which each copy is fired. Setting this value to 5 creates a copy every five frames. You will be using a cylinder as the object that you will emit. Remember when you create the cylinder to group it (if you are not using Sketchy Solids) so that you can apply the SketchyPhysics settings needed.

1. Open the SketchyPhysics UI, and make sure to also have the cylinder that you created selected.

2. Select the Emitter check box, type in a value of **100** for the Strength. Within the Lifetime box, type **200**, so that each copy of the cylinder will last for 200 frames. Finally, enter **5** for the rate value (Figure 10–22).

Figure 10–22. Emitter settings

3. Now run your simulation to see what happens (Figure 10–23).

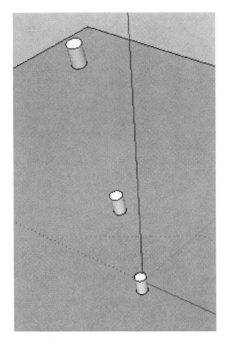

Figure 10–23. Simulating an automatic gun with an emitter in SketchyPhysics

Notice how copies of the cylinder are being fired. You might have noticed that
you need a little more control in the shooter. How can you do this? You can
control the firing process with the press of a key.

4.   Type in the following line of code into the Rate box:

```
key('s')*5
```

5.   Now when you run the simulation, the cylinder will not emit unless you press
the s key on your keyboard.

Animating the BA-64 Armored Car

Before concluding this chapter, you'll take a look at how to animate the BA-64B armored car you
modeled in Chapter 8.

The first step in animating the car is to create a solid floor on which the car will roll on.

1.   Click the "Create a floor" button.

2.   Then select File ➤ Import, browse to the Chapter 8 folder, and import the
wheel, chassis, and axle. You will have to import each part separately (Figure
10–24).

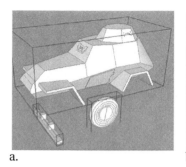 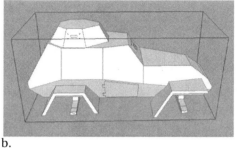

a.   b.

Figure 10–24. Chassis and wheel of BA-64

3.   Once the parts of the armored car are imported into SketchUp, make a copy of the axle, place one in front and one in the back, and group them with the body of the car (Figure 10–24b).

     Next you want to add the hinges to the wheel so the wheels rotate when you run the animation.

4.   Select the hinge, and click the center of the wheel to place the hinge, making sure that the circular rotation of the wheel is perpendicular to the wheel. Place the wheel next to one end of the axle, and then joint-connect the hinge with the body of the car. Then group the wheel with the hinge (Figure 10–25).

At this time, play the animation just to check whether there is a solid connection between the wheel and the car. Also make sure the wheel rotates as you move the model.

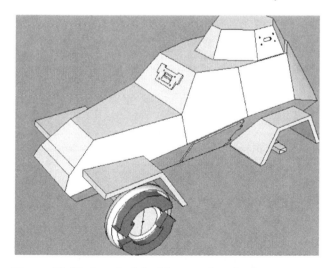

Figure 10–25. Attached the hinge to the wheel

5.   Now repeat step 4 for all the other wheels in the model.

Figure 10–26 shows what your model should look like once you are all done.

Figure 10–26. All four wheels attached

Now run your simulation; using your mouse, apply momentum to the car and watch it move. When you apply to much momentum to the armored car, it can be difficult to keep track of its location in the modeling window. For example, if you had created a huge landscape for the armored car to move on, it would move out of your perspective. Every time you would want to see the object, you would need to also drag the screen. This can be very frustrating, but there is a solution to this problem. You can set the camera to follow the model.

Camera Options

When the simulation is running, right-click the model, and you will be presented with camera options: Camera track, Camera follow, Camera clear, and Copy body (Table 10–8 and Figure 10–27). Select "Camera track," and the SketchUp camera will track the model.

Table 10–8. Camera Options

Name	Description
Camera track	The camera tracks the object.
Camera follow	The camera follows the object.
Camera clear	This clears the current camera settings.
Copy body	This copies the object that you have selected.

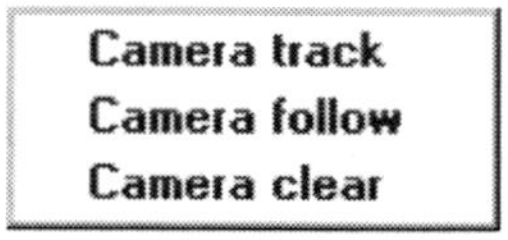

```
Camera track
Camera follow
Camera clear
```

Figure 10–27. Camera drop-down menu

Summary

I hope you found this chapter an interesting read. The goal of this chapter was to introduce a few of the tools available in SketchyPhysics that you can use to animate your models. It's not only a way to create some interesting animations but also a way to test the form, function, and performance of the model. And it's a great way to simulate a model before moving forward to 3D printing. If you are designing a model for a customer, having it animated can save you a lot of time. You can show how the model operates even before printing it. Both you and the customer will be happier at the end. Now that you are done with SketchyPhysics, in the next chapter you'll take a look at LayOut.

CHAPTER 11

■■■

Using LayOut

For any designer, presentation is important. The way you present your models will determine how others perceive your work. Good presentation leads to clarity; bad presentation leads to confusion. To assist you in presenting your models, SketchUp has introduced LayOut, which does things that other presentation software applications lack.

In this chapter, you'll use LayOut to prepare presentations. We kick things off with an introduction of the tools and panels in LayOut. You'll go through the basic steps of choosing a template, formatting the template, and adding text and images. You'll also import a model and add dimensions to it. In the concluding sections, you'll learn about annotating a LayOut presentation and how you can export your presentations to an image or PDF file format.

What Is LayOut?

Unlike animating your models, as you learned to do in Chapter 10, LayOut allows you to present your models like slide shows in PowerPoint, as well as create posters, banners, flyers, and much more. It's an excellent way for you to have a professional-looking printout of your model, without having to master advanced software packages such as Photoshop. And I love the fact that when edits are made to a model in SketchUp, they are automatically updated in LayOut. You don't have to spend time reimporting the model.

Getting Your Copy of LayOut

To get your own copy of LayOut, you will need to download the Pro version of Google SketchUp. You will have to dig deep into your pocket to purchase the Pro version, which currently sells for $495. If you're thinking of becoming a serious designer, then by all means purchase a copy. For those of you who are not sure about purchasing the Pro version, Google provides an eight-hour trial version of LayOut along with SketchUp Pro and Style Builder that you can download from the SketchUp web site. I recommend you try the trial version before investing any money. To download your own copy, visit the Google SketchUp download page at http://sketchup.google.com/download, and select Download SketchUp Pro.

After the eight hours are up, you can purchase a license from Google to activate the Pro version. Now, don't stop reading! There is great information in this chapter that I'm sure you will find useful. As I mentioned earlier, you will find features in LayOut that other presentation software applications lack.

LayOut Basics

Once you have the Pro version of Google SketchUp installed, you will see an additional set of icons. They should be located within the Programs folder under Google SketchUp; or check your desktop (Figure 11–1).

Google   **Style Builder**   **LayOut 3**
SketchUp 8   **2**

Figure 11–1. The Pro version comes with three software packages.

Double-click the LayOut 3 icon on your desktop. Notice that LayOut looks similar to other software packages. It has a menu bar (1), toolbar (2), document window (3), and panels on the right (4) (Figure 11–2).

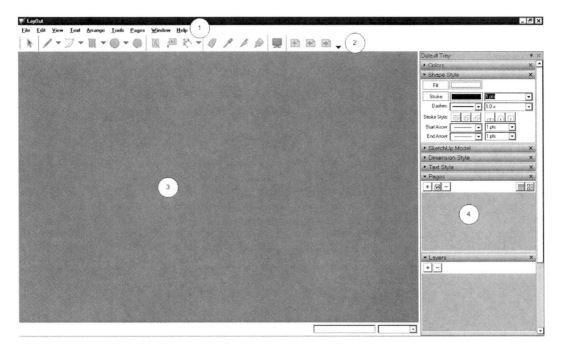

Figure 11–2. LayOut window

Before you get started using the software package, let's go through some of the basic tools available in LayOut to put together your presentations (Figure 11–3).

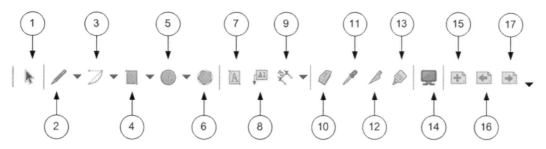

Figure 11–3. LayOut toolbar

There are 17 tools in the LayOut toolbar (numbered 1 to 17 from left to right in Figure 11–3). The function of each tool is described in Table 11–1.

Table 11–1. Selections in the LayOut Toolbar

Name	Description
1. Select	Selects objects within the LayOut window.
2. Line	Draws lines and shapes.
3. Arc	Sketches an arc.
4. Rectangle	Sketches a rectangle.
5. Circle	Sketches a circle.
6. Polygon	Draws a polygon. Type the number of sides followed by **s**, and press Enter.
7. Text	Inserts a text box within LayOut.
8. Label	Applies labels to models.
9. Dimensions	Adds text measurements to your model.
10. Eraser	Erases material from the LayOut window.
11. Style	Selects this icon, and your cursor will change to a dropper. Click an object in the model to select its style. The cursor will then change into a paint bucket. Select an object to apply the style.
12. Split	Splits a single line into 2.
13. Join	Joins lines together.
14. Start Presentation	Plays the presentation in full-screen mode.
15. Add	Creates additional LayOut pages.
16. Previous	Goes backward through LayOut pages.
17. Next	Goes to the next page in LayOut.

LayOut's Nine Panels

Most of LayOut's functionality can be found in the nine panels on the right of LayOut. In order from the top, they are as follows:

- Colors
- Shape Style
- SketchUp Model
- Text Style
- Pages
- Layers
- Scrapbooks
- Instructor
- Dimension Style

To display the contents within each panel, click the panel title. The Colors panel is the first of nine in the default tray menu (Figure 11–4). Select different colors to add to your presentation or document.

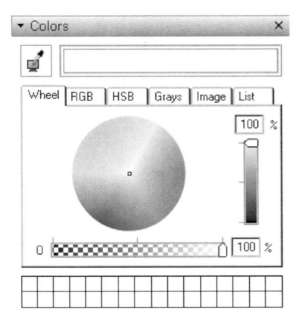

Figure 11–4. Colors panel

The second panel you will find is the Shape Style panel (Figure 11–5). With the Shape Style panel, you can format lines, arcs, and different shapes in the LayOut document window with fills, strokes, and dashes. Also, choose stroke style and arrow type.

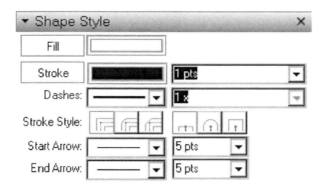

Figure 11–5. Shape Style panel

The SketchUp Model panel allows you to present your models in different forms, apply shadows or fog to your model, and orient your model to present different views (Figure 11–6). On the Styles tab, you can choose from an assortment of styles. The great thing about the SketchUp Model panel is with it you do not have to open SketchUp to edit your models. We will go through an example demonstrating how you can use this panel later in this chapter.

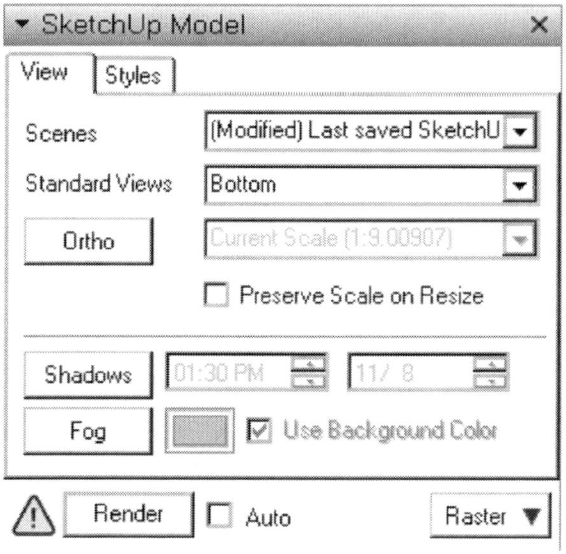

Figure 11–6. SketchUp Model panel

If you're looking to add a little flair to your model, then check out the Text Style panel (Figure 11–7). Here you can choose from a variety of fonts, typefaces, and sizes. You will be using the Text Style panel in this chapter to format text.

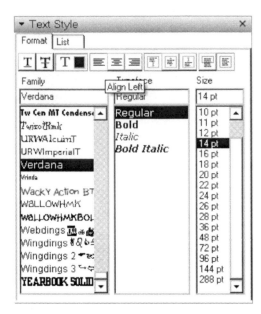

Figure 11–7. Text Style panel

The Pages panel indicates the number of pages in your layout (Figure 11–8). You can add and delete pages by clicking the + and - icons. You can duplicate a page by selecting the "Duplicate selected page" icon. Select the screen icon to the right of each page to toggle between either placing or not placing the page in your presentation. There is also an icon of a screen on the LayOut toolbar. Don't confuse that icon with the one here. The toolbar screen icon is for activating presentation mode.

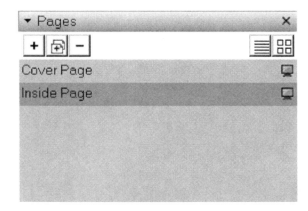

Figure 11–8. Pages panel

The Layers panel shows the number of unique layers within each page (Figure 11–9). You can add, delete, hide, and lock different layers. You will learn how all this works with the example in this chapter.

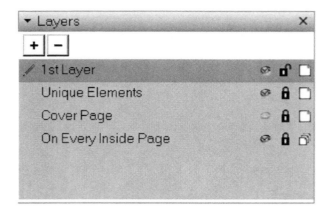

Figure 11–9. Layers panel

The Scrapbooks panel is a template in and of itself (Figure 11–10). You can store all kinds of material in a scrapbook and use it later. Place all the material (text, images, models) you would like to appear in your scrapbook into a single page. Then select File ➤ Save as Scrapbook. Enter a name for your scrapbook. Within the Scrapbook panel, select the scrapbook you just saved. Now you can drag and drop designs from your scrapbook into the document window on any page or document in LayOut. It saves you time not having to reimport material.

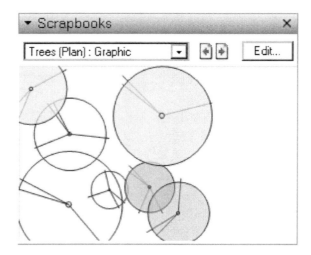

Figure 11–10. Scrapbooks panel

If you're new to LayOut and just getting comfortable with the software package, then your best bet is to keep the Instructor panel open. The Instructor panel animates the function of each tool within LayOut (Figure 11–11). Click the Eraser tool to see a demonstration of how the Eraser works within the Instructor panel.

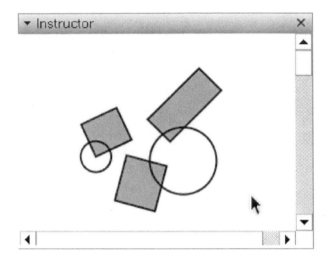

Figure 11–11. *Instructor panel*

Use the Dimension Style panel to adjust how dimensions you add appear in the document window (Figure 11–12). Select the demension you would like to adjust, and then within Dimension Style adjust its properties.

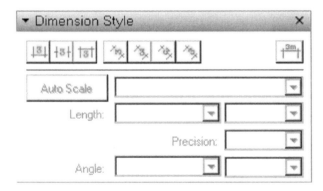

Figure 11–12. *Dimension Style panel*

Now that you have gotten a feel of all the tools and panels in LayOut, you are going to use some of these tools and panels to create a presentation.

Selecting a Template and Importing a Logo

In this section, you first will select a template and then import a logo to customize the template. At the end, you will import a model into your template and annotate it using the tools in LayOut. Along the way, you will learn how layers work and pages are created. By the end of this section, you will be a presentation pro.

Open LayOut, and the Getting Started dialog box will appear. You can also open the Getting Started dialog box by selecting File ➤ New (Figure 11–13).

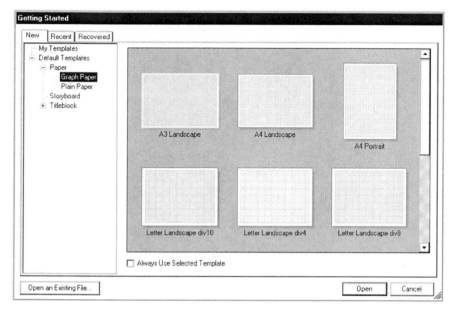

Figure 11–13. *Getting Started dialog box*

1. Within the Getting Started dialog box, select the New tab.

2. In the left column below the tab, you can select from an assortment of templates. Select Titleblock and then Rounded. On the right, select A3 Landscape.

3. On the bottom-right corner, click Open. The template will then open in LayOut and appear within the document window, as shown in Figure 11–14.

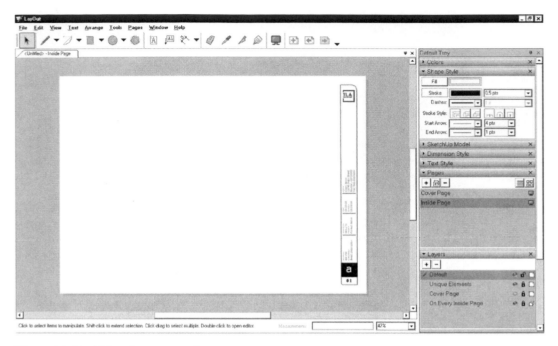

Figure 11–14. *A3 Landscape template*

You can find information about the template on the right under the default tray (Figure 11–14). Select Pages, and within the Pages panel are two pages—a cover and an inside page. These are the pages that currently make up the template. Depending on how many pages make up your presentation or document, you can add, remove, or rearrange the pages by clicking + within the Pages panel. You can also view the pages as a thumbnail or in list view by selecting buttons on the top right of the Pages panel.

Within the default tray, there is also the Layers panel. You will be using the Layers panel a lot for editing different parts of your template. You can use layers to divide different layers of your document or presentation. On one layer, you can place text and on another layer place your models. So, what's the point of doing all this? Well, if you have a large document and want to only work on editing the models in your presentation, you can edit it through a layer without having to worry about accidentally deleting any of the text.

Select the Layers panel, and see how many layers make up your template. Like the Pages panel, you can rearrange, remove, or delete layers. The icons on the right of each layer allow you to toggle the visibility of the layer, lock/unlock the layer, or share the layer across all pages.

To edit text and objects within the template, you will need to unlock the layer. To unlock a layer, click the lock symbol to the right of each layer. To lock the layer, click the lock symbol once again. After you have unlocked the layer, double-click the text within the template you want to change and type in the new text. As an example, you will go ahead and change the logo within the template. To do this, follow these steps:

4. Unlock the On Every Inside Page layer, and select the inside page in the Pages panel. Unlocking On Every Inside Page gives access to the title block.

5. Select the TLA logo on the upper-right corner of the template, and press Delete on your keyboard.

6. From the File menu, select Insert, and browse to the Chapter 11 folder; then select the image file titled House Model.png.

   You can download example files for this book from the book's catalog pages on Apress.com web site. Look on the catalog page for a section entitled Book Resources, which you should find under the cover image. Click the Source Code link in that section to download the image.

7. Select the logo, and then select Open.

Once you have placed the image onto your template, you will need to readjust the size of the image. To do this, click the image, and drag the blue triangles on the edges to shrink or enlarge the image (Figure 11–15).

Figure 11–15. *House model logo resizing*

8. Drag and place the logo in the location of the TLA logo you deleted in step 2.

9. Now unlock the Cover Page layer, and select the cover page within the Pages panel. Select and delete the TLA logo on the cover page, and then repeat steps 3 to 5 to place another copy of the house model logo.

Once that's all done, lock the cover page and On Every Inside Page layer. It's a good idea at this time to also save the template. Select File ➤ Save. You will be using this template as the default template for your designs. Remember to give the template a name, and save it in a location where you can easily access the file.

Understanding Layers

Understanding how layers in LayOut operate will make you efficient at developing documents and presentations. And they help you avoid the frustration of having to reenter text or models that you might have accidentally deleted in LayOut.

Within LayOut, there are two types of layers: shared and unshared. Changes made in a shared layer are displayed across all pages in a LayOut document. Changes made within an unshared layers are present only on the selected page in LayOut. To get a better understanding of shared and unshared layers, let's look at an example.

1. With the template you were working on in the previous section, click the + icon, and add a new layer. Double-click and rename the layer House Picture, and then select Cover Page (Figure 11–16).

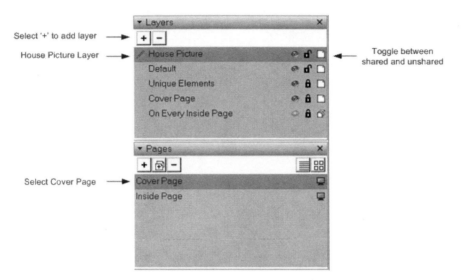

Select '+' to add layer

House Picture Layer

Toggle between
shared and unshared

Select Cover Page

Figure 11–16. Layers and Pages panel

2. Select File ➤ Insert. Browse to the Chapter 11 folder, and open the image file titled House Picture.png. Place the image in the upper-left corner of the cover page (Figure 11–17a). Now toggle back and forth between the cover page and inside page. Notice that the cover page only displays the house picture. For the picture to display on the inside page, you will need to share the House Picture layer.

3. Click the Page icon on the right of the House Picture layer, and then select the inside page. Notice this time the image also appears on the inside page as well (Figure 11–17b).

4. If you want the model to appear only on the inside page, select the page, and then click the page icon once more. The Unshared Layer dialog box appears (Figure 11–17c). Select the "Keep contents on this page only," and then click OK. Now toggle back and forth between the cover page and inside page, and notice that this time the image is only on the Inside page.

Using layers is very useful. When sections of presentation are divided into layers, it's easy to edit.

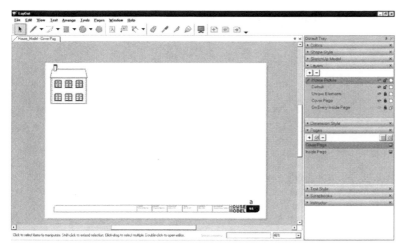

a.

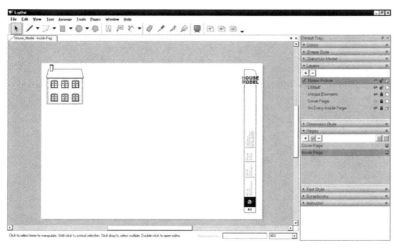

b.

c.

Figure 11–17. (a) Cover page; (b) inside page; (c) Unshare Layer dialog box

Now, using the same template, let's place a model of the house on both pages. But before you continue, make sure the House Picture layer is a shared layer and that the house image is displayed on both pages.

Importing a Model

To import a model of the house, you will use another layer. This way, you won't be inferring with the other material in your layout window.

1. Click + to add a new layer, and then name the layer Model or rename the Default layer. Lock all the other layers in the Layers panel.

2. To place a model into the template, select Insert ➤ File menu. Browse to the SketchUp file you want to insert. For illustration, you will be using a house model as an example. Within the Chapter 11 folder, select the house model, and then click Open. Place the model on the cover page.

    You can download example files for this book from the book's catalog pages on Apress.com web site. Look on the catalog page for a section entitled Book Resources, which you should find under the cover image. Click the Source Code link in that section to download the model.

3. When first inserting the model, notice that it is not correctly adjusted to fit on the page (Figure 11–18a). To adjust the dimensions, simply click the model. A blue box surrounding the model will appear with blue triangles on each corner. Move your cursor over the blue triangles, and then click and drag the triangle to resize the model. Also click and drag the model so it is in the middle of the page.

4. When a yellow triangle with an exclamation appears, you will need to rerender your model. This will occur when you adjust a model in LayOut. Right-click the yellow triangle, and select Render Models on Page (Figure 11–18b).

Having to rerender your model multiple times can be cumbersome. To automatically render the model, select Edit ➤ Preferences ➤ General, and select the check box titled "Automatically re-render SketchUp models as needed" (Figure 11–18c).

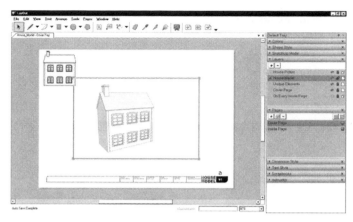

a.

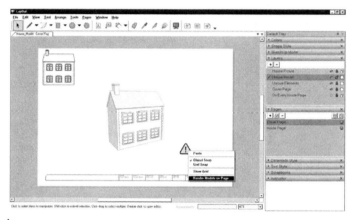

b.

c.

Figure 11–18. Placing and rendering a model on the cover page

5. When everything is aligned, select the inside page. Here you will be inserting the house model twice.

6. Select File ➤ Insert, and insert the house model twice into the inside page.

7. Move a copy of the model to the upper-right corner and the other copy to the lower-left corner of the page. Resize the dimensions of both models so that they take up only half of the page (Figure 11–19).

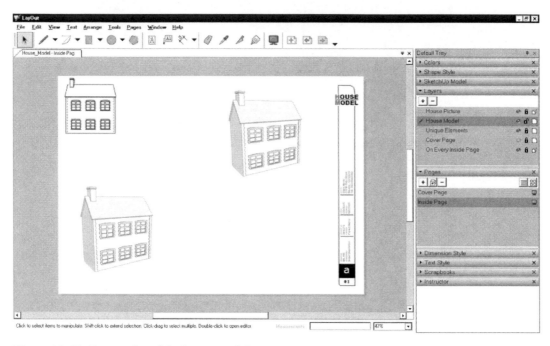

Figure 11–19. Two copies of the house model

8. Now with two copies, let's change the view of the model on the upper right of the page to see how you can easily change the view of the model without having to reopen SketchUp and adjust the model. Select the SketchUp Model panel, and click the View tab (Figure 11–20a).

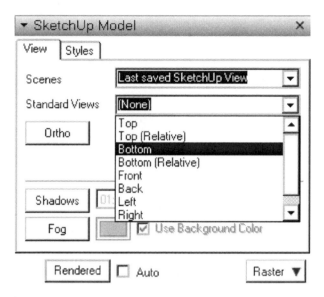

a.

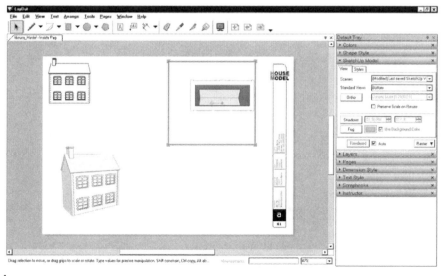

b.

Figure 11–20. Adjusting the models view in LayOut

9. Select the model on the upper right, and then from the Standard Views drop-down menu select Bottom. Now you should see the bottom of the house. It's really easy to change a view. You don't even need to open SketchUp. This is what makes LayOut great.

Adding Text to Your Presentation

So far, you have worked with inserting models and images, but what about adding text in LayOut? Actually, the process is very easy. LayOut provides two options: Text and Label. You will be applying text and labels to the house model you have been working with so far.

1. Before you get started, select the inside page.

2. Select the model in the lower left of the page. In the SketchUp Model panel, select the View tab, and in Standard Views, select Right.

3. Now you are ready to place text and labels. Select the Text button, click below the top model, and type **Bottom View**. Click above below the bottom model, and type **Right View** (Figure 11–21). If the text appears small, click the text, and then in the Text Style panel enlarge the font size.

4. You now will be placing dimensions to describe the width and height of the model. To place dimensions, select the Dimensions tool. Click the bottom corner of the Bottom View model, and then click once more on the opposite corner of the model. Drag the cursor out to place the dimension.

5. Using the same method, draw the dimensions for the other sides of the model. Repeat the process for the right view model. In case the text appears small for each dimension, select the text, and in the Text Style panel change the font size.

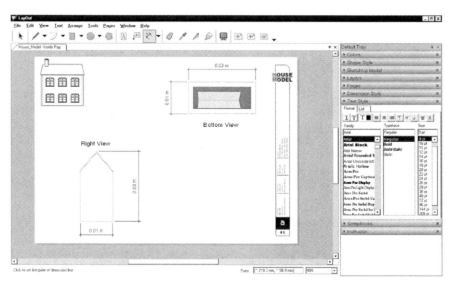

Figure 11–21. Bottom and right views with dimensions added for the house model

Masking

Sometimes the models you are working with can be complicated. With complex models, even greater complications can arise when the model is not presented effectively. How about describing part of the model that is difficult to see? One method of resolving this problem is masking certain parts of your model. Let's go back to the previous example you were working with and create a presentation that emphasizes certain parts of the model.

1. Select a shape to draw—this can be any shape from the toolbar. In Figure 11–22, I'm drawing a circle. Draw the shape over the part of the model that you will mask.

2. Holding down the Shift key, select both the model and the shape. With both selected, right-click, and select Create Clipping Mask. Now you have an isolated picture highlighting an area of your model (Figure 11–22). If at a later date you wanted to edit the location of the mask, then you can simply double-click the cropped image. Using the Select tool, you can move the shape.

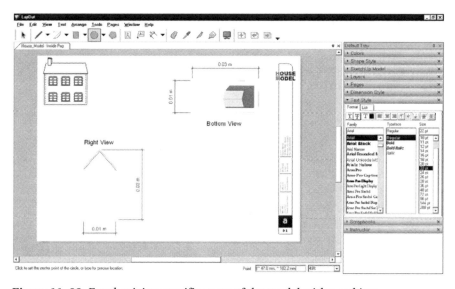

Figure 11–22. Emphasizing specific areas of the model with masking

Presentation

So far, you have gone through some of the basics of using LayOut, but there is actually a lot more you have not seen. Now that you have set up the design of the house, the next step is to present the model. You basically will be running your presentation in full-screen mode and annotating the presentation with markups in LayOut. To do this, follow these steps:

1. Click Start Presentation on the toolbar, denoted by the screen symbol. This will run your presentation, and you will see it on the entire screen of your computer Don't confuse the screen symbol in the Page panel with the one in the toolbar. Selecting the screen in the toolbar will place you in presentation mode. To transition through each slide, use the arrow keys on your computer. The right arrow key is for forward, and the left arrow key goes backward.

2. In presentation mode, notice that your cursor has changed into a pencil. The purpose of this pencil is to annotate your models. Click, and cross out the two middle windows in the model (Figure 11–23). Annotations allow you to emphasis different parts of your presentation so that your audience will better understand what it is you're talking about. You can also use it to mark up parts of presentation. As an example, I have crossed out the windows to let me know that the client would like it removed from the model. During a presentation, keep your annotations to a minimum. Only emphasize what is important. Don't over-clutter your presentation with too many annotations.

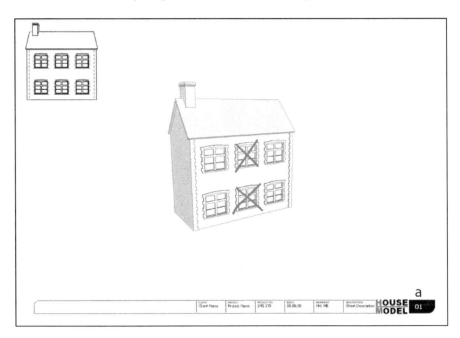

Figure 11–23. Adding annotations to your presentation

3. After crossing out the windows, press Escape on your keyboard. A dialog box will appear asking whether you want to keep the annotations or delete them (Figure 11–24a). Click Yes to keep the annotations; you can always remove them at a later in the document window (Figure 11–24b).

4. If you were presenting your models to clients, you would follow steps 1 to 3 to run and show your presentation. Also, as you read in step 2, you can annotate the presentation for your client.

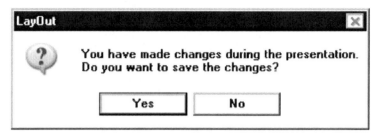

a.

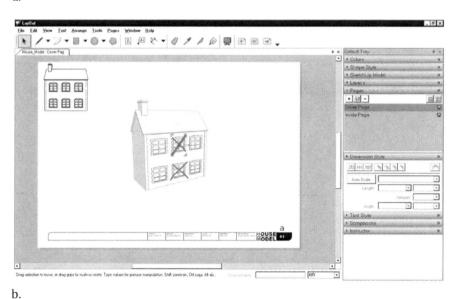

b.

Figure 11–24. Placing the annoation in your LayOut document

Exporting from LayOut to an Image or PDF File

The great thing about LayOut is that you can export the model either as an image or as a PDF file. Exporting your model in either format allows you to e-mail the design to the customer without them having to spend money to purchase the Pro version of SketchUp. Each page in LayOut is saved as individual image files. A PDF, on the other hand, stores the entire presentation on a single file. To export the presentation as an image or PDF file, follow these steps:

1. Select File ➤ Export and then Images or PDF.

2. A dialog box will appear asking for the location of where you want to save the image or PDF file. Select either .png or .jpeg for the image file. If you're planning to edit the model further, then it's best to save it as a .png file. The .png file format stores a layer of transparency into the model. If you saved the file as a PDF, you will need to download and install a copy of Adobe Acrobat Reader from http://get.adobe.com/reader/ to view the file.

LayOut Preferences

If you become an avid LayOut user, then the LayOut Preferences dialog box will surely come in handy, allowing you to set up parameters. To access the LayOut Preferences dialog box, select Edit ➤ Preferences (Figure 11–25).

Figure 11–25. *LayOut Preferences dialog box*

On the left of the dialog box is a list of eight preferences. Each is described in Table 11–2.

Table 11–2. LayOut Preferences Dialog Box Menu

Name	Description
Applications	With this option, you can select any application on your computer for editing images and text. Once an application is linked, you can access it within LayOut without having to browse through the Programs folder on your computer to access the program.
Backup	Graphic-processing software tends to crash more often because of the demands that it puts on the computer. It is always a good idea to have the autosave functionality selected.
Folders	This is the location where the templates and scrapbooks are saved.
General	This automatically rerenders a SketchUp model. Rendering generates an image out of a model.
Presentation	Select the monitor where you want to display the information.
Scales	Select among the available scales, or create your own custom scale.
Shortcuts	Select from an assortment of predefined keyboard shortcuts, or create your own keyboard shortcut to access the many options in LayOut.
Startup	Here you can set parameters when LayOut first opens. You can create a new document, reopen files from the last section, or not open anything. You can also select the types of document to open a blank or existing document.

For now I recommend you stay with the default LayOut Preferences settings. Once you are comfortable with LayOut, you will have a better idea of what options would be good to change.

Additional Tips

You can customize the look of the LayOut software by adding or removing tools from the menu bar.

1. At the corner of the menu bar, there is an arrow pointing down (Figure 11–26). Select Add or Remove Buttons, and then select Main Toolbar.

2. A drop-down menu will appear. Menu options that are currently active within the menu bar are indicated by a check. To remove a button from the menu bar, click one of the menu items. An item is removed when the check mark is not present.

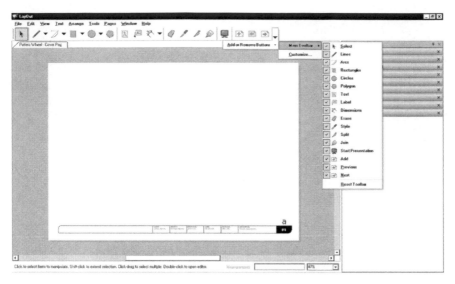

Figure 11–26. Displaying the toolbar

To add and remove panels from the default tray menu, follow these steps:

1. Select Window, and then deselect any of the panels that you do not want to use (Figure 11–27).

2. The Instructor panel is probably the one that you will use the least, once you are familiar with LayOut. You can go ahead and deselect the menu if you want.

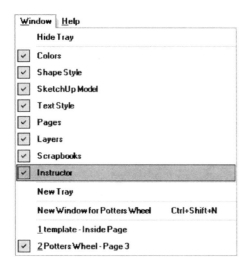

Figure 11–27. Window drop-down menu

Summary

In this chapter, you got an introduction of the tools and panels in LayOut. You went through the basic steps of choosing a template, formatting the template, and adding text and images. You also imported a model and added dimensions to it. In the concluding sections, you learned about annotating a LayOut presentation how to export your presentations to an image or PDF file.

LayOut is an amazing tool, and there is a lot you can do with it. LayOut provides features that are not available in many other software packages. With it, you can easily and effortlessly edit model styles and views without even opening SketchUp. Using LayOut, you avoid the frustration of having to reimport models, which saves you time.

CHAPTER 12

■ ■ ■

Exploring 3D Printing Alternatives

Throughout most of this book, you have spent a majority of your time learning about Shapeways and SketchUp. Both work well together when designing and developing models for 3D printing. The ease of using SketchUp and the ability to effortlessly upload and 3D print models are amazing. In this chapter, we switch gears and introduce a few other tools and services for designing and manufacturing your models. We start with an introduction to Ponoko and learn how to use this service for laser-cutting parts. We then briefly go over a few other 3D printing resources available online (Redeye, QuickARC, Xardas, and AlphaPrototypes). These services offer different material, apply different printing techniques, and convert SketchUp files to STL files compared to Shapeways. We conclude the chapter with a brief overview of the RepRap project and the fab@home machines.

Ponoko

Ponoko is an online manufacturing service to develop laser-cut parts and is similar in many ways to Shapeways. On its web site you can upload a model, order it, and have it at your doorsteps within no time. Unlike other manufacturing options, though, Ponoko specializes in laser-cutting and not 3D printing. If you are looking to develop two-dimensional parts, then Ponoko is the way to go. Ponoko's web platform is set up so you can also share and sell your own designs. Ponoko offers several different methods to upload designs onto its web site. We will go through an example demonstrating how you can use a SketchUp model and upload it to Ponoko for laser-cutting in the next section. You can use SketchUp and a camera or request a designer to design your product for you. This flexibility allows people of all levels to design. There are helpful tutorials on Ponoko's web site that will guide you through the process.

Ponoko also offers an assortment of materials for laser-cutting, as listed in Table 12–1; these are available at www.ponoko.com/make-and-sell/materials.

Table 12–1. Ponoko Laser-Cutting Materials

Name	Description
Corrugated card	Commonly known as cardboard, this material consists of three layers of paper—two outer layers and one corrugated inner layer.
Felt	This is nonwoven fabric made from compressed wool fibers.
Acrylic	This looks and feels like glass but is stronger. A common form of acrylic is Plexiglass.
PETG	Polyethylene Terephthalate Glycol (PETG) is commonly used in kitchen products such as Tupperware.
Bamboo	This is a fast-growing plant used for a variety of building projects.
MDF	Medium-density fiberboard (MDF) is made of compressed wood.
Veneer MDF	A thin layer of wood glued onto the surface of MDF gives the look and feel of a smooth surface.

Ponoko uses laser-cutters to manufacture its products. With a laser-cutter, you get cuts that are precise and dust free. The drawback is that they work only with 2D surfaces, and extra material is left over. The laser-cutter uses a computer-controlled beam of light to cut through the material. If you are using Ponoko, the size of the cutout has to be within 31.1 × 15.1 inches. For additional details, visit the Ponoko web site at `www.ponoko.com` (Figure 12–1).

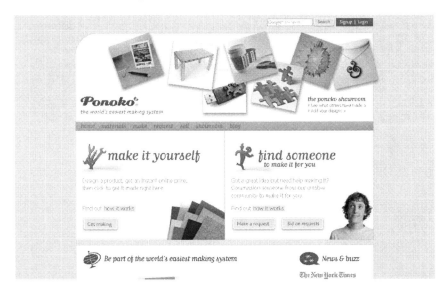

Figure 12–1. Ponoko web site

Before you continue to the next section, remember to create an account on Ponoko so you can upload you designs. To create an account, follow these steps:

1. On the Ponoko home page, click Signup in the upper right of the site. The signup page will appear.

2. Select between signing a free or a Prime account. The Prime account costs $39 monthly. Selecting either account will direct you to a form to fill in your name and address. Fill in the information, and agree to the terms or service.

3. Then click Free Sign Up! to process your account.

You are now done creating your account. Now we will go through the steps of preparing a design for upload.

Preparing Your Designs for Ponoko

In this section, we will go through the basic steps of preparing a Google SketchUp model for laser-cutting by Ponoko. To get started, you will need to download three items: the Scalable Vector Graphics (SVG) plug-in for Google SketchUp, Inkscape, and the Inkscape Startup Kit from Ponoko.

*   You can download the SVG plug-in from `http://code.google.com/p/sketchup-svg-outline-plugin/downloads/list`.

*   You can download a copy of Inkscape at `www.inkscape.org`.

*   You can download the Inkscape Starter Kit from the Ponoko web site at `www.ponoko.com/make-and-sell/downloads`.

▧ **Note** In case one or more of the links mentioned do not work, it's possible that the link has changed. Instead, use a search engine like Yahoo! or Google to search for the items.

If you are familiar with Adobe Illustrator CS, CorelDraw X3, or Macromedia FreeHand MX, you can use those tools instead of Inkscape. Unlike the other software packages, Inkscape is an open source program and is free for download (Figure 12–2).

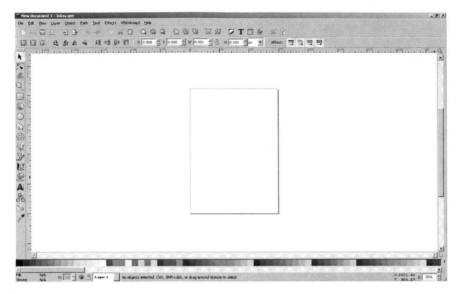

Figure 12–2. Inkscape

Once the SVG plug-in and Inkscape (or any other graphic editor) are installed, you are ready to go. Before you start to design your models, I recommend you first read the next section and understand the basic steps of preparing your model for development. You will notice that even though your model may look visually appealing and ready for development, there are usually some minor adjustments that need to be made before the model can be manufactured.

There are some design considerations you should prepare for before uploading your file to Ponoko. Remember that the maximum size of your model is limited to 31.1 × 15.1 inches and that Ponoko supports only the .eps and .svg file formats. Also, Ponoko will support a file upload of up to 10MB. Multiple files cannot have the same name, and all images must be converted to vector lines and fills before uploading to Ponoko. For further details, visit the Ponoko FAQ web site at www.ponoko.com/make-and-sell/designing-faqs.

Test-Tube Holder

In this section, you will be using an existing test-tube holder constructed in SketchUp as an example to be manufactured by Ponoko. Browse to the zipped file on the Apress web site for this book, open the Chapter 12 folder, and open the file titled Test Tube Holder. The design consists of six pieces: four legs, a bottom surface, and a top surface (Figure 12–3). You can also look at the model in Figure 12–3 and reconstruct it or design your own model to be laser-cut using Ponoko.

Figure 12–3. SketchUp model of test-tube holder

1. To export the file into Inkscape, select all the surfaces in the model by holding Ctrl on your keyboard and clicking to select each surface or by dragging the cursor around the entire model. For a model that has many parts, the second option is best.

2. Once the entire model is selected, right-click to bring up the contextual menu, and select Export to SVG. The SVG Export Preferences page will appear (Figure 12–4).

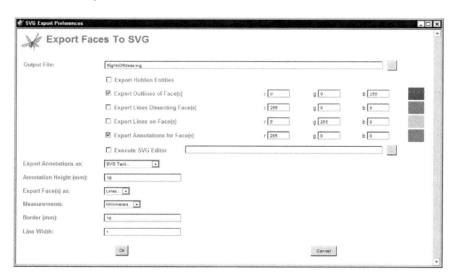

Figure 12–4. SVG Export Preferences dialog box

3. For the output file, select a directory where you want to save your files after export to SVG. When saving the file, give it the extension .svg, or it will be saved as an unknown file. Other programs you have used in this book automatically add an extension, but unfortunately this plug-in does not. Also remember to change your model's measurement to scale properly in millimeters or inches. Border allows you to set the distance between the selected surfaces.

4. Click OK to save your export preferences and create an .svg file for your model.

5. Open the .svg file that you just created in Inkscape or any other SVG image-editing software. You should see something similar to Figure 12–5a. If you selected the entire model, then you will see something similar to Figure 12–5b.

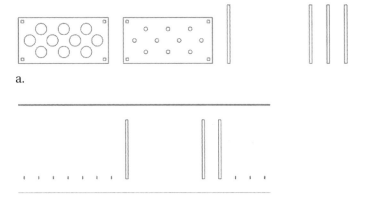

a.

b.

Figure 12–5. (a.) SVG image when individual files are selected; (b.) SVG image when the entire model is selected

Laying Out the Pieces and Uploading the Design

Now that you have all the pieces in Inkscape, you need to lay them out into one of the three Ponoko templates (the final design will then be converted into an .eps file). These templates have increasing safe-area dimensions to choose from and are located in the Ponoko Inkscape Starter Kit: P1 7.1" × 7.1" (181 × 181 mm), P2 15.1" × 15.1" (384 × 384 mm), and P3 31.1" × 15.1" (790 × 384mm), as shown in Figure 12–6.

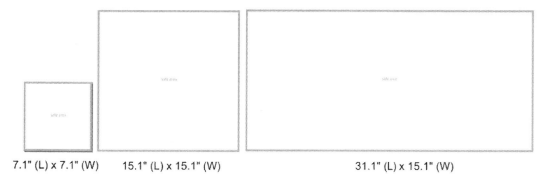

7.1" (L) x 7.1" (W)   15.1" (L) x 15.1" (W)   31.1" (L) x 15.1" (W)

Figure 12–6. Inkscape templates from Ponoko

1. Open the P3 template in Inkscape, and then copy/cut and paste the parts of the
    test-tube holder that you want cut out with the laser inside the orange box of
    the template.

2. After placing all the parts into the template, if there is extra space left over, try
    one of the smaller-sized templates. Make sure when you place the parts that
    they are not overlapping with each other. Inside the orange box is the safe area;
    anything placed outside will not be cut (Figure 12–7).

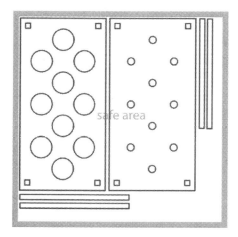

Figure 12–7. Ponoko 7.1" × 7.1" template

3. Once everything is placed in the proper location, export the file in .eps format.
    From the Inkscape File menu, select Save As, and save the file as an .eps file
    (Figure 12–8).

Figure 12–8. *Saving the file as* `.eps`

The design is now ready to be uploaded to Ponoko.

4. Log in to your Ponoko account, and upload the design.

5. If you have not created an account yet, select Signup on the upper-right corner of the Ponoko home page (Figure 12–9). The Signup page will appear. Follow the instructions on the site to guide you in creating an account.

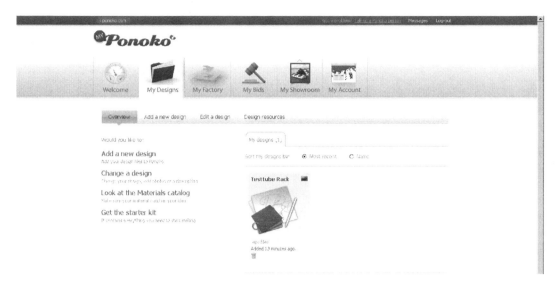

Figure 12–9. File uploaded onto Ponoko web site

Selecting the Cutting Material and Selling Your Design

Once you have uploaded a model into Ponoko, the next step is to choose the material on which the parts will be cut. Ponoko offers card, fabric, metal, plastic, rubber, and wood as options. Table 12–1 earlier in the chapter described the characteristics of some of the material. Some material is more expensive than others. Try to find a material that best suits the application of the model you have designed. For the test-tube holder, a plastic or wood material is a good option. They are both cheap and lightweight.

You are now ready to sell the design.

1. Click the Welcome button located on the top of the web site.

2. Then on the Welcome page, click "Sell a design/product."

3. Follow the instructions to set up a store to sell your design.

That's it—you are all done. I hope you found this section to be an interesting read. Now you can design models for 3D printing using Shapeways or use Ponoko to create 2D laser cutouts of your model. Before jumping to the next section, check out the Ponoko "showroom." Here you will find a gallery of Ponoko models that others have uploaded.

Exploring 3D Printing Alternatives

In this section, we'll briefly go over some alternative 3D printing services you can use to develop your 3D models: Redeye, QuickARC, Xardas, and AlphaPrototypes. Although this is just a small sampling, you can use them as an alternative to Shapeways. Read through this section, and browse through the web sites of each service. Some services apply different printing methods and use different material. See whether any of the services would be a good source for your 3D printing needs.

Redeye

Redeye is an online rapid prototyping service. If you're looking to build professional designs, this is the place for you. One great thing about Redeye is that, like Shapeways, it provides an instant quote for your model without having to wait. The file format that Redeye accepts is STL. For more details about the Redeye service, visit www.redeyeondemand.com (Figure 12–10).

Figure 12–10. Redeye web site

QuickARC

Rather than converting your files into STL format, you can submit your 3D model to QuickARC to convert the design for you (Figure 12–11). Its focus is on 3D printing architectural models. QuickARC offers coloring of your model, and for printing it uses plastic powder, liquid plastic, ABS plastic, nylon plastic powder, and acrylic-based polymer. QuickARC accepts the STL file format for 3D printing, but it also supports SketchUp files and automatically converts them for you.

Figure 12–11. QuickARC web site

Xardas

Xardas (Figure 12–12) uses fused deposition molding and ABS plastic to manufacture models. Fused deposition molding is a form of 3D printing that applies an additive process to deposit material layer by layer through an extrusion nozzle. Xardas accepts models that have a maximum allowable size of 10 × 10 × 12 inches. There is an assortment of colors to choose from for your finished model: green, yellow, white, black, red, and gray. Xardas uses the STL file format but also converts any file format to STL for you. There are some predesigned models that you can download from Xardas, but the selection of models is limited compared to what Ponoko and Shapeways have to offer.

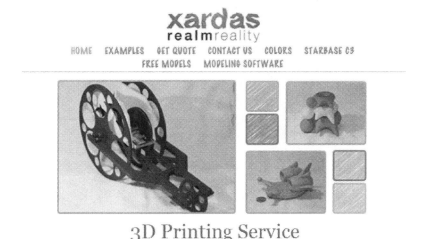

Figure 12–12. Xardas web site

AlphaPrototypes

AlphaPrototypes is similar in many ways to Xardas and QuickARC, because it is a rapid prototyping company specializing in printing 3D models. It uses ABS and composite materials for printing. To order custom-made parts from AlphaPrototypes, the CAD file has to be in the STL format. Once uploaded, the models are instantly price-quoted. There are four printing methods to choose from with AlphaPrototypes: FDM, PolyJet, SLA, and Zcorp. There are pros and cons to each printing method. More information and a comparison of strength, accuracy, pricing, colors, and fine detail can be found on the AlphaPrototypes web site at www.alphaprotypes.com (Figure 12–13). In addition to many of the other prototyping services, AlphaPrototypes offers the option of painting, sanding, priming, and plating your design with copper or chrome.

Figure 12–13. AlphaPrototypes web site

When submitting STL files for prototyping to AlphaPrototypes, the file sizes need to be between 5MB and 10MB. Also, keep in mind if you are working with AlphaPrototypes that there is a $149 minimum order.

Personal 3D Printers

3D printers have been available for many decades but were accessible only to universities, research institutes, and industry. New advancements in technology and a relative reduction in cost have allowed many consumers to now obtain a personal 3D printer. These days, you can purchase a good-quality 3D printer for less than $10,000, which is relatively cheap compared to 3D printers 15 to 20 years ago. In this section, you'll explore a couple personal 3D printers you can purchase for your home or small office.

V-Flash Desktop Factory

You can purchase the V-Flash Desktop Factory at www.desktopfactory.com (Figure 12–14). The V-Flash Desktop Factory can be easily placed on a desktop and hooked up to a laptop or desktop computer. Because of its compact size measuring only 26 × 27 × 31 inches, with very little effort the printer can be placed anywhere in your home or office. The printer weighs 145 pounds and can build models that are 9 × 6 ¾ × 8 inches in dimension. This would be a great investment for a small business or for a school.

Figure 12–14. Desktop Factory web site

The file type supported by the printer is STL. Your SketchUp files can easily be converted to .stl files using MeshLab, an open source 3D model–editing software (http://meshlab.sourceforge.net/), and then uploaded to the V-Flash for 3D printing. If you don't have a couple of grand to spend on V-Flash, then your best bet would be to purchase the CupCake CNC Starter Kit (Figure 12–15). For just $649, you can have your very own 3D printer.

Figure 12–15. The MakerBot store

Purchasing the kit requires you to have some technical troubleshooting skills, so take a look at the setup documentation before purchasing the product, and see whether this is a project you can conquer (http://wiki.makerbot.com/cupcake). Estimate spending about a week getting the kit up and running. It takes a day or two to construct all its parts and a few days troubleshooting and making sure the material extruder works correctly.

RepRap

For the techies out there, an even ambitious alternative to purchasing a 3D printer is constructing one from scratch. If you enjoy assembling, tinkering, and immersing yourself in an activity where you get to energize your brain cells, then the project to get involved in is the design of the RepRap. The CupCake CNC is actually an offshoot of the RepRap. The development of the RepRap was started by Dr. Adrian Bowyer at the University of Bath in the United Kingdom. His ambitious goal was to develop a machine that could mimic the replicating abilities of plants and animals. In this fashion, the RepRap was designed to print most of its own parts. Since the start of the RepRap in 2005, it has seen several makeovers. The project saw huge growth with the development of the Darwin in 2008. By 2009, the second generation of RepRap came out called the Mendel. The current model called the Huxley is similar to Mendel but is smaller in size.

The RepRap is an open source project that allows anyone to share and modify the design of the project without any patent restrictions. Four parts make up the design of the RepRap: extruder, XYZ platform with motors, electronics, and software. Most of the parts for the development can be purchased online. You can find details on the RepRap web site at www.reprap.org (Figure 12–16).

Figure 12–16. RepRap web site

If at any time you get lost and are not sure what to do, visit the large community base of RepRap users. Visit the community portal to find out developments of the RepRap and meet other RepRapers at http://reprap.org/wiki/RepRapWiki:Community. Post your questions on the RepRap community site (http://reprap.org /forums.reprap.org), or read one of the many blogs on the development of the RepRap.

fab@home

A slightly different but similar project to the RepRap is the fab@home 3D printer. Like the RepRap, it is an open source desktop 3D printer. Development parts can be easily purchased online. There are currently two versions of the printer: Model 1 and Model 2. Because of its plastic design, the machine is a little more costly than the RepRap. Expect to spend close to $2,000, which is still better than purchasing a commercial one.

What's unique about the printer are the syringes. The syringes come in two models and can be used with an assortment of materials: thermo plastic, thermoset, electrically conductive, and ceramic materials. For more details on each type of material and to learn more about the Fab@home project, visit www.fabathome.org (Figure 12–17) and the wiki page at www.fabathome.org/wiki for a getting-started guide to the project.

Figure 12–17. fab@home web site

Summary

Wow! What an adventure. You finally made it to the end of this book. Congratulations! In this chapter, you looked at a few alternatives to 3D printing your models with Shapeways. The chapter started with an introduction to Ponoko for laser-cutting SketchUp models. Then you looked at a few alternative sites where you can send your models off for 3D printing, and we discussed some of the features they have to offer. At the end of the chapter, you saw a couple personal 3D printers you can purchase or build from scratch.

Now that you are done reading the book, what's next? Check out the appendix, where I introduce you to ways in which you can connect with the 3D modeling and printing community.

■■■

Get Connected

This appendix is all about connecting you to the 3D printing community. Here you'll learn about online forms, blogs, digital fabrication sites, and some plug-ins you might want to check out as your adventure continues beyond this book in 3D printing and modeling.

Community

Do you want to share your modeling expertise, learn from other modelers, find answers to difficult problems you're facing while modeling, and help those who can benefit from your struggles? The best way to do all of this is to join an online community. Both Shapeways and Google SketchUp provide a community base with a huge collection of resources.

Shapeways Community

The Shapeways Community link gives you access to the Shapeways blog, forum, Shapeways live chat, events, newsletter archive, testimonials, and contests. The Community link is located on the menu bar of Shapeways. Clicking the link will direct you to the Community page (Figure A–1). Table A–1 describes what you would find if you were to select any of the links.

Figure A–1. Shapeways Community page

Table A–1. Links Under Community

Feature	Description
Blog	Visit the blog to read about recent developments at Shapeways, work from other 3D designers, fairs, contents, and general news about the 3D printing community.
Forum	If you're trying to find a solution to a Shapeways problem, search the forums or post a question. Or share your experience with someone in need of help with Shapeways or 3D modeling.
Shapeways Live Chat	Sign up to be reminded of the next live Shapeways webcast, or log in to the chat room to talk with other Shapeways users.
Events	Learn about the upcoming Shapeways events.
Newsletter Archive	Sign up to receive newsletters from Shapeways.
Testimonials	Post a testimonial about your experience with Shapeways.
Contests	Looking for a challenge? Join a Shapeways modeling contest.

Google SketchUp Community

The Google SketchUp community has been around for a lot longer than Shapeways and has gone through a lot of versions in the process (see Figure A–2). So, I'm sure that if you face any difficulty while using SketchUp, you can easily find a solution in the help forum, blogs, or newsletters.

Figure A–2. Google SketchUp community site

You can access the Google Community page at http://sketchup.google.com/community/. The site has an abundance of information, which is briefly described in Table A–2.

Table A–2. Google SketchUp Community Site

Name	Description
Case Studies	Learn about what some of the SketchUp Pro users have been developing. Find case studies in the Architecture and Design, Digital Entertainment, Construction & Engineering, and Education categories.
Gallery	Take a look at images of SketchUp models.
SketchupUpdate Newsletter	Receive monthly updates of the best post on the SketchUp blog.
Press	This area is all about Google SketchUp in the news.

Name	Description
Resources	Find training, plug-ins, books, and more to further develop your SketchUp talents.
Developers	Interested in developing plug-ins for SketchUp? Click the Developers link to learn about the SketchUp Ruby API.
Google SketchUp Help Form	Can't find an answer to your problem? Post a message, and someone from the community will give you a hand.
Google SketchUp Blog	Read about contents, community events, and recent developments of SketchUp.
Go Green with SketchUp	Learn how you can use SketchUp for energy analysis.
SketchUp Pro for Non-Profits	Learn how you can get SketchUp Pro as a nonprofit.
Project Spectrum	Learn how SketchUp is helping the autism community.

Additional Blogs and Sites

The Shapeways and Google SketchUp pages are great resources, but there is a lot more that you can learn. Here are a list of few sites I recommend you visit to learn more about the world of 3D printing and modeling:

- Cnc Zone (www.cnczone.com). Join a community of designers and learn about CNC machines, CAD/CAM software, and how to buy and sell CNC machines.

- Fabbalo (www.fabbaloo.com). This is a blog all about fabrication, desktop manufacturing, and 3D printing.

- Instructables (www.instructables.com). This is a web site dedicated to all things homemade, handmade, and do-it-yourself. Type in the words **3D printer** and **CNC** to see what others have built and how they have built them.

- Replicator (www.replicatorinc.com). This blog written by Joseph Flaherty has all kinds of information about the world of digital fabrication. His posts are all about recent developments in 3D printing, new designs, new technology, companies, and products.

- Build Your Own Cnc (www.buildyourcnc.com). This is a site about the development of a personal CNC machine by Patrick Hood-Daniels. He provides step-by-step video instructions for designing your own CNC machine from scratch.

- Meetup (www.meetup.com). This is an online repository of groups you can search in your local area or around the world. Do a search of the words *SketchUp, Shapeways, 3D modeling,* or *fabrication* to see whether there is anyone else within your community that you get together with and share your passions.

Additional Plug-ins

In this book I covered a few plug-ins in Chapter 5 and SketchyPhysics in Chapter 10. There is an endless number of plug-ins online for Google SketchUp. It would require at least another three to four books to cover them all. And new ones are being developed all the time. Here are a few additional plug-ins worth checking out:

- SU Podium (www.suplugins.com). Developed by Cadalog, Inc., this is a photorealistic rendering plug-in. Now you can take models you have designed in Google SketchUp and make them look like real photographs.

- Slicer (www.cad-addict.com/2009/07/sketchup-plugins-slice-your-model.html). This is a great plug-in developed by TIG for 3D designers and printers. The plug-in lets you take slices of your model. The Slicer will go a step further and place all the slices on a single plane. Using an SVG exporter, you can print the design on paper or send it to them for laser cutting on plastic.

- Waybe (waybe.weebly.com). This is a plug-in for unfolding your SketchUp models. Print cutouts of your model that you can then fold and hold in your hands.

Index

■ M

LaVergne, TN USA
16 December 2010

208945LV00006B/57-70/P